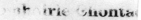

SCIENCE, TECHNOLOGY
AND CULTURE

■ I S S U E S ■ in CULTURAL and MEDIA STUDIES

Series editor: Stuart Allan

Published titles

SCIENCE, TECHNOLOGY AND CULTURE

David Bell

OPEN UNIVERSITY PRESS

Open University Press
McGraw-Hill Education
McGraw-Hill House
Shoppenhangers Road
Maidenhead
Berkshire
England
SL6 2QL

email: enquiries@openup.co.uk
world wide web: www.openup.co.uk

and Two Penn Plaza, New York, NY 10121–2289, USA

First published 2006

A catalogue record of this book is available from the British Library

ISBN-10: 0 335 21326 X (pb) 0 335 21327 8 (hb)
ISBN-13: 978 0335 21326 9 (pb) 978 0335 21327 6 (hb)

Library of Congress Cataloging-in-Publication Data
CIP data applied for

Typeset by RefineCatch Limited, Bungay, Suffolk
Printed in the UK by Bell & Bain Ltd, Glasgow

CONTENTS

SERIES EDITOR'S FOREWORD

'Scientists have finally broken the hound barrier,' my newspaper reliably informs me today, because they have succeeded in cloning an Afghan puppy called Snuppy from the skin cells of a three-year-old male dog. In heralding the arrival of Snuppy – short for the Seoul National University puppy – as the latest breakthrough in the fast-moving 'world of genetic manipulation', the news report declares that it brings to an end 'a seven-year worldwide race to replicate a dog from donor cells using the technique pioneered by British scientists when they cloned Dolly the sheep in 1998'. Another day, another scientific wonder. Across the mediascape, whether it's the daily news or the latest Hollywood science fiction blockbuster, science is seen to be providing answers, solving problems, and making the world a better place – at least, that is, when it's not threatening us with imminent extinction by spiralling out of control. One need not invoke a language of causative media 'effects' or 'impacts' to acknowledge the formative ways in which we draw upon media representations – factional and factual alike – to help us make sense of scientific controversies. And what of Snuppy in this regard? 'Sadly,' one scientist commented, 'the media interest is likely to attract pet owners keen to recreate their much-loved pets, although this demand is unlikely to be met until the efficiency of cloning is raised.' Meanwhile, she added, the cloning of animals 'raises many ethical and moral issues that have still to be properly debated within the profession'.

It is in seeking to contribute to new ways of thinking about these issues that David Bell's *Science, Technology and Culture* makes its intervention. As is made clear from the outset of the discussion, this book calls into question a range of familiar assumptions that typically underlie approaches which maintain that science and technology work 'outside' of culture, and thus simply produce effects on it which can be measured objectively. In thinking about science and technology as culture, Bell elaborates an alternative way of conceptualizing the complex ways in which their significance is negotiated – however unevenly – in everyday life. An array of fascinating examples are used to illuminate the contours of this new approach, including objects such as boiled eggs, refrigerators and 'gay brains', as well as media representations revolving around the portrayal of scientists in technoscience films, the technologicalization of music performance, and the symbolic depiction of Moon landings and nuclear conflagration, among others. Along the way, Bell encourages us to become properly self-reflexive about 'the tools we use to think about the tools we use' and, in so doing, shows us why a cultural studies of science and technology, 'with its scruffy mixture of theories and methods, and with its political commitment to understanding both the promises and threats of science and technology, refusing any one-size-fits-all perspective', has so much to offer us. This is a richly perceptive book, one that is sure to be welcomed by those up to the challenge of thinking anew about our current scientific and technological opportunities and predicaments alike.

The *Issues in Cultural and Media Studies* series aims to facilitate a diverse range of critical investigations into pressing questions considered to be central to current thinking and research. In light of the remarkable speed at which the conceptual agendas of cultural and media studies are changing, the series is committed to contributing to what is an on-going process of re-evaluation and critique. Each of the books is intended to provide a lively, innovative and comprehensive introduction to a specific topical issue from a fresh perspective. The reader is offered a thorough grounding in the most salient debates indicative of the book's subject, as well as important insights into how new modes of enquiry may be established for future explorations. Taken as a whole, then, the series is designed to cover the core components of cultural and media studies courses in an imaginatively distinctive and engaging manner.

Stuart Allan

ACKNOWLEDGEMENTS

Most of the ideas in this book were worked through first in the classroom – the cultural studies classroom at Staffordshire University, in the module Techno-cultures. That module, and the degree in cultural studies at that institution, no longer exists as a result of 'cut backs'. This book owes its greatest debt, there-fore, to the staff and students that passed through that particular moment in time, and hopefully learnt something along the way. Thanks to you all. The staff who helped me teach these ideas included Phil Cartledge, Andrew Conroy, Mark Featherstone, Mark Jayne, Barbara Kennedy and John O'Neill, and I owe them all much thanks. For reading and commenting on parts of the book as it came together, and for conversations about its themes, thanks to Jon Binnie, Tim Edensor, Ruth Holliday, Joanne Hollows, Mark Jayne, Martin Parker and Tracey Potts. And, of course, thanks to Stuart Allan, the series editor, and Chris Cudmore at Open University Press, for patience and for chivvying me along. Special thanks, as ever, to Daisy, Ruth, Mark and Jon.

David Bell

1 | SCIENCE, TECHNOLOGY . . . AND CULTURE?

Science and technology are too important to be left to scientists and technocrats themselves.

(Steven Best and Douglas Kellner)

There's an email that periodically does the rounds at work, usually just before the start of the academic year. It's aimed at lecturers, people like me, who are preparing for a new cohort of undergraduates to arrive. Purportedly originating from a college in Wisconsin, and updated or supplemented as it zaps from inbox to inbox across the **Internet**, it offers a series of cautious reminders about the life experiences of the next intake of students. While it makes me feel terribly old, I want to pick out some of its list here. About the students entering higher education today, it says:

- They have never feared a nuclear war.
- They are too young to remember the space shuttle blowing up.
- Atari predates them, as do vinyl albums (except as an elite product used by professional DJs). The expression 'You sound like a broken record' means nothing to them.
- Most have never had a reason to own a record player.
- They have probably never played Pacman or heard of Pong.
- The compact disc was introduced before they were born.
- They have always had an answering machine.
- Most have never seen a TV set with only four channels, nor have many seen a black-and-white TV.
- They think that portable DVD players are useful, there have always been VCRs, but they have no idea what BETA is.
- They cannot imagine not having a remote control.
- They were born after Sony introduced the Walkman.
- They find mobile phones absolutely essential for normal life, and have never

learnt how to plan ahead, because they can change all arrangements instantly with mobiles.
- They can't imagine what hard contact lenses are.
- There has always been MTV.
- The Titanic was found? They thought we always knew where it was.
- They don't have a clue how to use a typewriter.

Now, while there is an element of fogeyism at work here, and an awful lot of simplification and assumption about who these students are and what places and backgrounds they come from, there are also a lot of interesting things to unpack. The first is the rapid churning of new technologies, and their equally rapid domestication. It *is* hard to imagine life without remote controls or answering machines if you've never lived without them, just as it's hard to get used to them when they're newly introduced. And each of these bullet points has a story, or a series of stories, behind it – the marketplace battle over video cassette formats, in which VHS triumphed over Betamax (see Mackay 1997), or the history of the Sony Walkman (Du Gay *et al*. 1997), or the impact of MTV on reshaping the pop music industry (Kaplan 1987), or the rapid 'cultural obsolescence' of computer games (though these have now been given a second chance thanks to the vogue for retro-gaming; see King 2002). And each of these stories is about science and technology – about invention and innovation, research and development, microchips and the stuff of physics and chemistry and ICT lessons. But they're also all about culture, about the ways our lives have given shape to and been shaped by arcade games, personal stereos, remote controls (on the latter, see Michael 2000). These are the stories I want to tell in this book.

There are other stories to be told, too. At the heart of this book is the story implied by the title of this chapter: what does it mean to think of science and technology as culture – and in both senses: what does this mean for science and technology, and what does it mean for culture? What does it mean, also, for the study of science and technology, and for the study of culture? I want this book to contribute to these stories, just as the email quoted above has been added to over time, as it gets passed on in cyberspace. Storytelling is a very important business, in fact. It is one way of understanding what culture means: 'the stories of which we find ourselves a part', as John McLeod (1997: 27) nicely puts it. Perhaps that means it's time to try telling that part of the story – the culture bit.

The culture bit

Turn to any entry-level cultural studies student text, and you will find an attempt to define culture. Such definitions always seem to be both cumbersome and angst-ridden – that in itself being part of the story of cultural studies. They tend to remind us that Raymond Williams said that culture is a very complicated word, that it can mean something like civilization (in the sense of 'being cultured'), that it can refer to creative output (whether elite, high culture or mass, popular culture), and that it can mean ways of life. Where I used to work we used a four-part definition which I still quite like, noting that 'culture' can refer to *products* (sometimes called texts, even when they're paintings, cakes or personal stereos), *practices* (singing, gardening, texting, sometimes called lived cultures), *institutions* (museums, broadcasters, governments – bodies that produce and regulate what counts as culture) and *theories* (ways of understanding the other three; we might also include methods here, in terms of ways of finding things out about culture).

As you can see, the common device used in definition is exemplification. You will also have spotted how I've sneaked in science-and-technology examples in my own definitions above. Critics will say that the problem with this kind of expansive definition is that it means that *everything* is culture, so the term becomes too baggy, and hence meaningless. But I think that misses the point: if we agree that culture is what people make, or do, through their interactions with other people and with things, then it gets hard to find something that's truly 'outside' of culture. Aha, you might say – what about nature? But it is humans that have 'invented' nature, by calling some things natural, and other things variously 'man-made', or artificial, or technological, or cultural. Our very understanding of nature is therefore cultural, leading some writers to conjoin and coin words like '**naturecultures**' (Haraway 2003). OK, if nature is culture, you now say, then surely things like science can't be? Science isn't like soap opera – the usual cultural product used to criticize cultural studies. You can guess what's coming next: of course science is culture! Or, to use some cultural studies lingo, it is a set of signifying practices that produces, circulates and consumes texts, that interacts with 'non-science' in various ways, that has a range of (cultural) institutions associated with it, such as universities, and that produces very powerful stories and ways of storytelling. Next objection: science doesn't tell stories, it makes discoveries, produces theories and facts. A mindtrick I often use to work this through uses the idea of the locatedness or situatedness of culture; the idea that culture is embedded in place and time. So looking at science and technology in different times and places should help us get to grips with this cultural embeddedness. A nice little example of this is provided by the historian Thomas Lacquer (1990), in his book on the history of sex. The example concerns what he calls an 'interpretive chasm' about a story of sex and death. Let me recount it here.

The story concerns a young monk on his travels, who comes into a scene of grief at a country inn. He is told that the innkeeper's beautiful only daughter is dead, her funeral arranged for the following day. The innkeeper asks the monk to watch over her body for the night, which he agrees to do. Only trouble is, the talk of the dead young woman's great beauty gets the monk curious, and he looks upon her. She is, indeed, beautiful – so beautiful, and seemingly so dead, that he cannot resist her, and 'takes advantage' of her lifeless form. Horrified by this necrophilia, he hurriedly vanishes in the early hours. At the funeral next morning, just as the dead woman's coffin is about to be interred, movements and sounds are heard from within it – the mourners open the coffin to find the innkeeper's daughter alive, her 'death' having been what we would now call a coma. Later it becomes clear she is not only alive, but pregnant – though she has absolutely no idea how.

This tale is used by two eighteenth-century physicians, according to Lacquer, to reach very different conclusions, and then reinterpreted in the mid-nineteenth century, again to make a very different point. The first interpretation concerns death – the story is used to prove that only scientific tests can prove a person is dead. Sexual ecstasy could bring one out of a coma, but not back from the dead. But the monk swore that the 'dead' young woman showed no such signs of ecstasy. And this raises the fundamental problem for the second interpretation, which concerns sex. Without such ecstatic, orgasmic – and unmistakably 'lively' – reaction, the innkeeper's daughter could not possibly have become pregnant. As Lacquer (1990: 3) puts it, 'the girl *must* have shuddered, just a bit'. Hence the tale is one of deception – the monk and the innkeeper's daughter conspired to hide what had truly gone on in the inn that night. Revisited in 1836, the story is given a different medical spin: the woman was in a coma, the monk did get her pregnant, therefore female orgasm is unnecessary, even irrelevant, to conception – overturning eighteenth-century medical wisdom. This decoupling has far-reaching consequences, to this day; as Lacquer puts it, it meant that 'women's sexual nature could be redefined, debated, denied, or qualified. And so it was of course. Endlessly' (ibid.).

But, you might say, all this means that in the past people didn't know any better, that with advances in medical science the reproductive process has come to be understood more clearly. The path of science leads always towards the truth. But if we look back at those eighteenth-century tellings, and the later retellings, we can also see that what gets taken as true at any time is uncontested, but not uncontestable. The absolute necessity of female orgasm for conception gets replaced by its absolute irrelevance, but neither of these ideas is less true at the time it is taken as truth. Put simply, truth is cultural, too.

And while science has put a claim on truth in cultures like ours, we have to see it as only one way of thinking about what counts as truth (see Chapter 6).

Now, a recurring problem with this line of argument, and one used regularly to bash people trying to argue that science and technology are culture, is that to say there are different truths means everything and nothing is true, which means being unable to choose between truths, which means descent into chaos! This really is too big a debate to get into here; you are going to have to just run with me on it (either that, or put this book down now). To say that there are competing truth claims, and that science holds a good hand here and now, does not mean that, to use the catchphrase deployed to dismiss this line of thinking, 'anything goes'. But it does make it important to understand what does go, how it goes, who it goes for, and so on. History can help us with that, as I hope the example taken from Lacquer illustrates.

And so can geography.[1] So my second example is about how things are thought about differently in different places, and what that teaches us. Helen Verren's (2001) *Science and An African Logic* draws on her experiences in primary schools in Nigeria, where she saw first-hand a very different approach to numbers, mathematics and teaching. The Yoruba number system at the heart of her book works around a base of 20. I will need to quote here to make sure I get this right: 'Utilizing a secondary base of ten, and then a further subsidiary base of five, integers emerge. There are fifteen basic numerals from which the infinite series is derived' – in English these are one, two, three, four, five, six, seven, eight, nine, ten, 20, 30, 200, 400, 20,000 (Verren 2001: 55). Numbers in between are named by their relationship to these bases, so the number we call 50, called by the Yoruba *àádóta*, is derived thus: $(-10 + (20 \times 3))$. Larger numbers have more than one name, based on the different ways of using the number system to derive and thereby name them – so the number we call 19,669 has seven Yoruba names, for example. Now, Verren is keen to point to the politics at work here in the way this number system has been viewed and treated, both in terms of the effects of colonialism and postcolonialism, and in terms of the globalization of particular forms of numbering. She also focuses on the 'social life of numbers' – how the way we think about and use numbers has multiple effects on how we think about the world and our place within it. Mathematics, numbers, as you've no doubt guessed by now, are culture.

Part of the politics at work here, as already noted, is concerned with the replacement of such 'local' ways of knowing with those assumed to be 'universal'. As Verren points out, in the case of Yoruban numeracy, colonialism has a lot to answer for, in constructing alternative number systems as inferior, 'primitive'. This translates as: 'in different places they see things differently, but we know better because science transparently produces truth.' We'll see more of the politics of science, technology and culture throughout this book.

For now, I want to delve a bit deeper into the science-as-culture, technology-as-culture, pair of equations. Connecting back to my earlier discussion of what culture means, we need to note a number of ways in which science and technology might be thought of as cultural. If we start with some immediate ways, then move outwards, we can get a sense of the manifold ways we need to think these equations. So, to begin: I am sitting here, writing this, on a personal computer. It looks like a television tied to a typewriter, with a grey box stood nearby, and that funny little clicker, the mouse, at its side. In my lifetime, I have seen the computer enter into everyday life through a series of steps or stages; each step has been both technological and cultural (for a fuller account, see Bell 2001). That computers look like they do, work like they do, and evoke the feelings that they do – these are all cultural moments. We might think about the 'computing counterculture' and its ideas about democratic computing, which helped birth the laptop, the mouse, the home-build PC, the human–computer interface (HCI), and the whole language of usability. Or we might ponder science fiction's representations of computers, from Hal to the Demon Seed to SkyNet, and how those images are refracted through our daily interactions with computers near and remote. Or perhaps you should just look at that flickering screen looking back at you, look at its lights and buttons and icons, listen to its hum. In all these and countless other ways of thinking about computers, we must surely see culture writ large (on lots of these ideas, see Fuller 2003). As Ellen Ullman (1997: 89) writes, 'we place this small projection of ourselves all around us, and we make ourselves reliant on it'. Stop and think about that for a moment. A bit of autobiographical scrutiny always helps me (again, see Bell 2001) – so think about *your* computer, your practices of computing, all those taken-for-granted interactions and practices, from trusting a spell-checker on a word processing package, to sticking Post-Its around the monitor to remind yourself of Important Things To Do (see Fuller 2003; Lupton and Noble 1997, 2002).

To break it down a bit more, we can begin by thinking about computers as *things*, as technological objects or artefacts. As we'll see in Chapter 3, there's a move towards making sure that people who think about culture or society don't forget about things, don't only think about people. But what kind of thing is a computer? We can look at it as a commodity – it has a price and a value, is bought and sold, and its status as a commodity allows it to accrete certain other meanings and associations, for example through the language of advertising. Go into a computer superstore and look for yourself. But, as people who study this kind of thing have shown, we feel ambivalent about things-as-commodities, and although we like the showing-off that comes with buying new and expensive things, we also need to decommodify them, to make them 'mine' (Appadurai 1986). In fact, things pass through a number of life stages, as they move from commodity to prized possession, to everyday object and then to obsolescence, and computers are no exception. In many ways they are an amplified, speeded-up version of this, given the rapid redundancy of new technologies (itself part of a bigger story, one of unending refinement and infinite replaceability; see Lehtonen 2003). Once away from the store, welcomed into our lives, woven into routine, the computer loses its commodity veneer (though we are reminded of its worth, for example by insurance and warranties, and by the threat of theft). It is, in the parlance, *domesticated* (Bakardjieva 2005; Lally 2002; Lehtonen 2003). We customize it, fill its hard drive up with our stuff, buy add-ons, tinker with screen savers and wallpaper. We start to live with it. (Of course, the way this works out depends on where we locate the computer – at home, which room, at work, or in the non-place of the laptop.)

So, now it's not so obviously a commodity, what kind of thing has it become? It's still a technological object, though as Timothy Taylor (2001) reminds us, the '**technologicalness**' also recedes as we learn the techniques of use (see also Chapter 3). It becomes an everyday object, its identity defined by its use as much as anything else. It's now a work tool, a communications device, a toy, a prosthetic memory, or a bad reminder of an unwise impulse buy. It still has some 'commodityness', too, as in the forms already mentioned, plus, for lots of us, its continued drain on our finances through credit agreements, broadband rental, and so on. Now, as Kevin Hetherington (2004) has written so evocatively, things in our everyday lives kind of disappear over time. They get pushed aside by newer, shinier things; or the everyday repetition of use makes them habitual, and thus unnoticed. In the case of the computer at home, it becomes a carrier or container of other things, its 'computeriness' receding. It can be a constant reminder of work, of pressing deadlines; it can be the source of household conflict over priority use; it can become equivalent to other domestic objects, especially now, when computers can also be televisions, MP3 players, sex aids, movie studios, and so on.

The backgrounded home computer can be brought back into the limelight, of course, and we can be re-reminded of our intimate, even codependent relationship with it. Nothing brings this out more sharply than a broken or stolen computer. A broken computer reminds us of another way of thinking about the kind of thing a PC is: it is a **black box** (and yes, I know it's grey, or taupe). Black boxing is explained in Chapter 3, but let me rehearse it a little here: a black box is a device full of inner workings that it is users don't need to know or understand. It is enough to know what goes in and what comes out. A lot of technological artefacts are black boxes for most of their users. This brings its own anxiety-effects, of course, and its own dependencies – on helplines and users' manuals, or the 'warm expertise' of more computer-literature friends (Lehtonen 2003). For most of us, our personal computer is heavily black boxed.

We know how to turn it on, how to get it to do some of the things we need, how to get help when we're lost. But how it *works*? As we shall see later, and in Chapter 6, knowing a bit about how things work is important, because it helps us connect back out to the cultural questions: why does it work like that, what are the impacts of it working like that, and so on.

Let's recap. The personal computer can be thought of as a commodity, as an everyday object, as a black box. It can end its days as a redundant piece of junk, a 'bygone object', to be disposed of somehow, either by literally throwing it away, or by resale, or by passing it on to someone else, or stashing it somewhere out of sight (Lehtnonen 2003). Christine Finn (2001) has written a brilliant book on computers-as-junk, in which she looks at the ways they are disposed of, all the activities that take place at the supposed end of a PC's days, whether that means having its reusable bits removed, or being snapped up by a vintage computer collector. People from my generation don't like to think about computers as junk, because to us they're still such new things. I find it much harder to throw one out than, say, a tumble drier, and I can see much more (symbolic) value in a 20-year-old computer than I can in a 20-year-old car.

The trade in retro or vintage computers also shows us how computers work as memory objects (see also Bell 2004b; Laing 2004) – another way of thinking about them. Then there are the myriad 'subcultures' that have sprung up around computing, from gamers to **ASCII** artists, from hackers to webcam voyeurs (Bell 2001). The list goes on . . . As you can see hopefully, computers are technological *and* cultural things, through and through; they always have been, as their history teaches us (Ceruzzi 1998). And they always will be, judging by the proliferation of uses to which they may now be put, both intended and unintended (on the latter, the so-called **'double life' of technology**, see Chapter 3). The computer is a *cultural text* or *product*, in other words. Or, more accurately, it is a set of texts, both hardware and software, written (encoded) and read (decoded) in heterogeneous cultural contexts. It also generates and participates in a range of *cultural practices* along with humans and other things (practices like writing this book, or selling old computers on eBay, among millions of others). It has produced a set of *cultural institutions* (including multinational corporations, virtual communities, the aforementioned eBay, or MIT's famous Home Brew Computer Club). And to help us understand it, we can turn to growing bodies of *cultural theory*, some pre-existing and borrowed to help think things through, some born out of computer culture itself, as in Fuller's (2003) discussion of 'software criticism' (see also Bell 2001).

A good example of this kind of many-headed approach to thinking about technology as culture is provided by Jonathan Sterne (1999), who offers up a 'manifesto' for how cultural studies might study the Internet (obviously this is both less than and more than the computer culture I have been sketching; see also Bakardjieva 2005). This involves moving beyond the hype, whether positive or negative, about cyberspace; 'mundanizing' or 'ordinari-izing' the Internet by showing how it works in everyday life and by treating it more as other media and technologies are treated. Sterne also calls for a historical, sociological, empirical and theoretical account of the Internet that can 'offer an effective critique of existing discourses around the medium, present some effective tools for thinking about it, and even provide a cogent discussion of its future' (Sterne 1999: 281). To my mind, although it sets itself less grandiose aims, Danny Miller and Don Slater's (2000) *The Internet: an Ethnographic Approach*, based on fieldwork in Trinidad, comes close to at least some of Sterne's manifesto calls. In particular, in working with everyday users of the Internet, it strikes a move away from discourse to practice (although acknowledging the inter-weaving of the two; when people use the Internet they are always interacting with the discourses around it at the same time; see Bell 2001).

Of course, as Sterne's article suggests, science and technology isn't just objects, it's also knowledges, ways of thinking. This has been an area of real conflict between people who think scientific and technological thinking is different, meaning special, and those who want to show that it cannot be insulated from culture, or the social, no matter how much this is claimed. In social and cultural studies of science and technology, discussed in Chapters 2 and 3, we will see how these so-called **Science Wars** have been framed, and what kinds of arguments are fired back and forth. And then there are the practices that are performed by those doing science and technology. As sociologists of science have shown, the daily science work that takes place in laboratories, for example, gives rich insights into these performances, with their power plays, their subplots and denouements. But labs are black boxes, too, for most of us – things go in them (people, ideas, matter of differing kinds) and out comes some new science or technology, whether a theory or a gizmo. Outside of school, we don't get many chances to go inside the laboratory, other than through its fictional representations (see Chapter 4). We aren't expected to understand what goes on in real laboratories – partly *because* we've seen their cinematic counterparts, all bubbling test tubes and arcs of electricity, and so expect real labs to be that simple and that exciting. (Of course, the reverse isn't held to be true: everyone should be able to understand things like cultural studies, even if only to joke about it; see Warner 2002.) As we'll see in Chapters 4 and 6, the idea of the **public understanding of science** is important if ambivalent, even controversial.

In place of knowing what goes on in day-to-day science and technology work, then, how are non-scientists supposed to think and feel about it? We are expected to *trust* it. We are expected to be able to differentiate the wild fictions of sci-fi from the truth of science and technology as tools for humankind's benefit. We are expected to support science and technology, to value it, to welcome its interventions into our lives. But sociologists have shown that *risk* is the flipside of trust (Beck 1992; Giddens 1991), that science fiction and fact are less separable than is suggested above, that people see the dangers as well as the promises of science and technology, that they need reassurance that it will produce benefits rather than monsters. The same science that took people to the Moon also threatened to destroy us all with the Bomb (see Chapter 5) – and anyway, what's the point of going to the Moon, when science can't solve more earthly problems? People are confused, uncertain, suspicious. One way that science and technology institutions have sought to remedy this is through the project of popularization. This involves trying to make non-scientists better understand that science is their friend, and teaching us how to tell good science from fraudulent pseudoscience (see Chapter 6). Popularization is attempted, for example, by staging people-friendly displays of science, whether in whiz-bang interactive science and technology centres (ISTCs), or in popular TV shows or websites (Fahey *et al.* 2005). It is also attempted by writing books that explain science through the lens of ordinary people's ordinary lives.

Boiled eggs, visible humans and gay brains

One such attempt is Len Fisher's (2002) *How to Dunk a Doughnut: the Science of Everyday Life*, a snappy little book of anecdote, experiment and explanation that belongs to a long tradition of using 'ordinary' things to explain core scientific ideas. It includes discussions of the science of biscuit dunking, throwing and catching, the bubbles in beer and in the bath, and my personal favourite – 'How does a scientist boil an egg?'[2] The answer is, of course, *perfectly*, because he or she understands the physics and chemistry behind this everyday act. A graph shows that the egg white cools as it sets, because setting needs energy. And Fisher explains that the white sets at a temperature lower than the yolk, because

> the protein molecules in the yolk are each wrapped around a tiny core of oil. It takes more energy to release the protein from the oil surface than it does to unwind an albumin molecule in the aqueous environment of the white – the yolk proteins are not free to move around and become entangled until the yolk reaches a higher temperature than the white. The yolk, in fact, only sets above a temperature of 68C, so the problem of boiling an egg becomes a matter of getting the white above 63C [at which temperature it sets], while keeping the yolk below 68C.
>
> (Fisher 2002: 39)

Fisher suggests ways to make egg boiling more accurate, by making it more scientific – while also admitting that 'gastronomic experience' can also produce reliable effects.[3] What's going on in this account, and indeed throughout Fisher's book and others like it, is one variety of science popularization. By showing the science behind boiling an egg, people might become curious about the science behind other everyday acts, like dunking a biscuit (there's a great diagram of the optimum angle a biscuit should be dunked into a hot drink, accompanied by an explanation of diffusion). This might make them think more positively about science. At the same time, the book does try to offer science-based advice, such as how to calibrate temperature and timing to get a perfect soft-boiled egg.

Not everyone thinks this is the best way to go about things. As we will see in Chapter 6, there's a slide in some critics' assessment from popular to populist, bringing in the horrors of 'dumbing down'. Fisher's book is and isn't dumb; it's also funny, charming, scientific, pedagogic, pedantic. One of the objections to this form of popularization is that it makes science into a sideshow attraction, trivializing it. Now, this objection depends on something quite unscientific, I think – the *a priori* prediction of who will read it and how it will be read. Scientific knowledge is not just produced, it is also consumed, by other scientists and by non-scientists. Often this process is mediated by intermediaries, 'translators' who repackage science for its public (Allan 2002). At other times we encounter it directly, for example by picking up Fisher's book – though here he presumes the level of translation needed to make it comprehensible. But with the opening up of newer, less regulated channels of communication and dissemination, the act of translation becomes much more difficult. The Internet represents probably the most contested site of scientific knowledge for exactly this reason.

An example of this second, ambivalent form of popularization-in-cyberspace can be found in the form of the Visible Human Project (see Waldby 2000). Launched in 1994, the VHP comes from the US National Library of Medicine, and marks an attempt to use high-powered computation and imaging technologies to disseminate online a digital archive of the human body. It does this by creating complete, anatomically-detailed, three-dimensional virtual renderings of two human corpses, one male and one female. These images can be looked at, manipulated, animated, revivified and vivisected from anywhere, for ever. With applications in medical research, surgical training and biomedicine pedagogy, the VHP has also attracted a lot of popular attention; there are some amazing animations, 'flythroughs' and so on available online, some of dubious humour. Catherine Waldby's (2000) excellent book on the VHP describes the production, circulation and consumption of these uncanny cyber-zombies, endlessly available, infinitely reproducible; their flesh made into data, frozen in time and scattered across cyberspace. As a project in science popularization, the VHP has to contend with its own ghoulishness, including the handling of the stories behind the people whose corpses became the Visible Man and Woman. The animations accessible online, especially the flythroughs, provide an unprecedented level of (mediated) access to human anatomy for most non-medical users, including those who teach and learn cultural studies. One 'double lesson' it teaches us, as Waldby says, is that science and technology's representations of the human body in fact reveal its *hold* over the body, provoking a new wave of Frankensteinian fears about the 'digital uncanny'.

Biomedical mapping of human anatomy has a long and problematic history, described by Michel Foucault (1975) among others. But it also occupies a central role in science popularization, with 'popular anatomy' being a cornerstone of science education and 'popular science culture' (along with space, dinosaurs and computers). The human body can be experienced as a black box by many of us; we know what goes in and comes out, but very little about how all the guts and goo do their jobs. The VHP at least makes the body's interior morphology available to us, in glorious Technicolor, helping us see what a leg muscle looks like, or how the duodenum works. You might even be able, with a trained eye, to pick out a particularly controversial bit of anatomy that I want to talk about here in the context of the complex popularization effects of science and technology: the interstitial nucleus of the anterior hypothalamus (INAH). In 1991 an American neuro-anatomist, Simon LeVay, wrote about an observed difference in a particular neuron group in some dead brains, some from homosexual men who had died with AIDS, the rest from heterosexual men and women, some of whom had died with AIDS (LeVay 1993; Murphy 1997). He found variance in one nucleus, INAH3, in part of the brain thought to be responsible for assorted so-called instinctual drives, including some around sexual activity. IHAH3 was found to be smaller in the 'gay brains' (as they would become short-handed in subsequent press coverage). While rejecting the claim that this proved a biological basis for homosexuality, LeVay did say that his findings suggested that 'gay men simply don't have the brain cells to be attracted to women' (cited in Murphy 1997: 27).

LeVay's research belongs to a field concerned with exploring the biological basis of sexual orientation, itself part of a broader field which has considerable popular appeal, and which looks for biological explanations (and increasingly genetic ones) for aspects of human behaviour. In his useful book on sexual orientation science, Timothy Murphy (1997: 25) writes that there is a pressing need to understand how this research 'passes into and shapes public consciousness'. While wanting to defend it as an area of legitimate scientific enquiry, Murphy asks that this is not presumed to be sealed off from culture, from the uses to which such research might be put. While some 'sexual minorities' enthusiastically received the scientific proof of their sexual orientation, which might enable them to make rights claims on the basis of discrimination, others feared the possible eugenic consequences, not to mention the impact of 'hypo-thalamic essentialism' on long-fought arguments about the social construction of sexuality (Bell and Binnie 2000; Tiefer 1995). As Murphy concludes, the fears that sexual orientation science provoke – fears of the 'science running amok' variety common to contemporary genetic anxieties (Turney 1998) – show the profound split in contemporary culture between science and ethics, with scientific research prompting both simplistic interpretations about 'gay brains' and also anxieties about 'science brains' – a point we will explore in much greater detail in the next chapter. This does not mean, as we'll see, that science and technology work 'outside' of culture and simply produce effects in it. The desire to research sexual orientation in this way emerges from and is shaped by cultural issues, too.[4] Having dissected visible humans and gay brains, as well as having learnt how to boil an egg scientifically, I now want to sketch in brief the contents of the rest of *Science, Technology and Culture*.

The anatomy of this book

Having mapped out some of my key concerns and issues in this introduction, and hopefully convinced you that there are useful ways of thinking about science and technology as culture, the next two chapters work at deepening this understanding. Although it results in an untidy split that I will attempt to suture at the end of Chapter 3, in these chapters I will initially consider science and culture, then technology and culture. The task of both sections is to point up some of the most important and interesting arguments from a range of perspectives, in order to give some further substance to the issue at the heart of this book.

Chapter 2 surveys selected moments in social and cultural studies of science, mapping the development of the histories and sociologies of science in their various forms. Key players such as Kuhn, Merton and Popper are name-checked on the way, and the idea about science proceeding as 'punctuated equilibrium' is introduced. At the heart of this discussion is getting to think about how science comes to have the authority and veracity it does, how it makes truth-claims (partly by marking other things out as non-truths; see also Chapter 6). In showing what social and cultural studies of science have to offer in unpicking and unpacking the cultural work of science, Chapter 2 shows how researchers have sought to defamiliarize science, for example by studying it in an anthropo-logical way, or writing about it as if it's a docudrama, or even as if it's being studied by Martians. In 'weirding' science, these strategies ask us to look anew at how science works. Strategies just like these have proved very unpopular among some segments of the science populace, sparking the so-called Science Wars, a spat over the legitimacy of science studies and the legitimacy of science.

This chapter is followed by its twin, which provides a route-map of social and cultural studies of technology. It starts with a fridge, as a nice, everyday way of pondering the shape and meanings of technology and how we might think about it as culture. Although the fridge is classed as a 'white good', and even though the one I talk about is silvery, we can see fridges as black boxes. This chapter starts to explain this important concept; then, like its predecessor, it introduces key bodies of thought, including those that explore the social shaping of technology or the **social construction of technology** – and here we take a bike ride and also have a brief look at **actor-network theory**, meeting fuel cells and scallops and trying to understand what happens when we try to find a way to talk about humans and non-humans and the practices of 'heterogeneous engineering'. The chapter then turns its attention to ideas about technology and everyday life, sketching some studies about domestic technologies and household gender politics, before looking at how these might play out in the future, as imagined through the idea of the **smart house** (such futurological work is quite important in parts of this book). Finally, it points to critics who argue that splitting science off from technology has become meaningless (thanks, guys), and who instead write about the conjoined '**technoscience**'. There are important reasons behind this, which the end of Chapter 3 tries to explain.

Then it's time to watch some films and listen to some music, in an effort to think about how popular culture represents science and technology. Understanding the cultural work of representation is very important to us here, since most non-scientists rely upon sites such as the cinema to help them think about their relationship to science and technology. The chapter addresses only two realms of popular culture in any detail – film and music – but you could easily add your own. The discussion is partly framed by a debate about the public understanding of science, and spends a lot of time looking at filmic images of scientists, often shown to be lab-coated boffins lost in their work and lacking social skills (and sometimes social conscience). The impact of these images should not, of course, be simply read off: audiences occupy various reading positions, and bring to the cinema all kinds of ways of thinking about science, technology and movies. Academics also have lots of ways of thinking about these things, and Chapter 4 scrolls through some branches of film theory, showing how they have been or might be used to unpack technoscience films. Then the chapter moves to consider popular music as a site where different science-technology-culture work takes place. It shows how the technologicalization of music performance and production sparks anxieties about musicianship or musicality, and thus valorizes the idea of 'liveness'. It shows how some music plays with ideas of technology, and it shows how a whole range of technologies are involved in the 'circuit of music culture', from microphones to MP3 players, and from mouth organs to MTV.

Chapter 5 takes a different tack, in looking at two particular symbols of the intertwining of science, technology and culture. It is the most personal chapter of the book, unapologetically so; it discusses the Moon landings and 'nuclear paranoia', emblematized in the Bomb. Flying to the Moon, and Mutually Assured Destruction – these are, for me, the most potent symbols of the promises and perils of science and technology. They are both, of course, intensely cultural symbols, too. This chapter tries to produce a 'cultural reading' of all these things at once, to think about how the Moon and the Bomb still cast shadows over our ways of thinking about science, technology and culture.

The last substantive chapter considers the boundaries of science and tech-nology, and the 'boundary work' that keeps some things in and other things out of the science club. It introduces CSI-COP, one of the sci-fi sounding bodies who patrol the borders of science, looking for hucksters and charlatans. It begins in the bookshop, browsing the shelves of *Science Of . . .* books, but also taking a sneaky peek at UFO sightings atlases. For a moment in this chapter I attempt yet more splitting, this time by trying to make a distinction between fringe and pseudoscience. Realizing the folly of my ways, the chapter moves on to talk through examples of both, asking what connects and separates them as bodies of knowledge and practice. Among its stop-offs are New Agers, cars running on water, low-carb diets, astrology and ufology. The point here is to think about what's at stake in erecting and policing exclusionary boundaries. The history of science shows time and again that fringe ideas can become new orthodoxies, as we shall see in Chapter 2.

Finally, at the end, there are a few last thoughts. I want to end this introduc-tion with one of these, a thought prompted first by the email with which I opened this chapter, and reinforced by reading books on science, technology and culture written only a few years ago. This final note of worry is about *ephemerality*. I'm reminded of the churn rate in all three of the things this book is about; and perhaps especially in their intersections. Does anyone remember *The X-Files*, for example? That show made such a splash, but is now relegated to endless rerun on specialist cable channels. Things change, but that doesn't stop them from mattering. So please forgive any embarrassing anachronisms here (the Beagle 2 landings – who remembers?). If we agree with Bruno Latour (1993) that *We Have Never Been Modern*, then you can't really expect one little book to be *that* modern, either, can you?

Notes

1 This is the first time I will mention geography – as you'll see in Chapter 2, it occupies a special place in my own tale of science, technology and culture.
2 This sounds as if it should be a joke, but I've been unable to come up with a suitable punchline. Suggestions? And, of course, it raises the prospect of a follow-up joke: how does a cultural critic boil an egg? Again, punchlines, please.
3 Not so if you believe a report in the UK's *Waitrose Food Illustrated* magazine in April 2005, in which top chefs were asked for tips on boiling eggs, which were then tested back at the mag. The differing advice, and the different outcomes, suggest that even considerable 'gastronomic experience' can't produce good eggs.
4 Both LeVay and Murphy locate themselves as gay-identified scientists, their research motivated at least in part because of their experiences as gay men.

2 | THINKING ABOUT SCIENCE AND CULTURE

Scholars who link science to culture are motivated . . . not by a desire to discredit science, but to understand it in a new way. We want to know science in a way that some scientists might not want to hear about, as a concrete social practice carried on by real people in a world of values, interests, influences, and drives.

(W.J.T. Mitchell)

When I first went to university, in the early 1980s, it was to study for a joint degree – a Bachelor of Science, no less – in geography and geology. (I had really wanted to do just geology, but wasn't deemed good enough at science – but that's another story.) Recently, I was moving boxes full of old books, including those from my student days, and I came across one volume that started me thinking about my degree studies, thinking about the two subjects I was immersed in, thinking about the different ways they were taught and learnt. Now, I don't want to claim any universal generalizations from my experiences, just to use them as a springboard into yet more thinking – thinking about science and culture, or science *as* culture.

The book that catalysed my nostalgic moment was Arild Holt-Jensen's (1980) *Geography: Its History and Concepts*. The book reminded me of how excited I was as a first-year student, listening to some introductory lectures that talked through some of the themes of that book: stories of how different ways of thinking and doing geography came about, who thought them up, and what became of them. Even in a straightforward linear narrative, the stories were thrilling. They were, as I remember them, stories about *human* geography – an important distinction I shall return to in a moment. The main events included the quantitative revolution and spatial science, Marxist geography, and so on. Something about the ways in which human geography had worked over time, absorbing and chewing over different ideas, different perspectives, different theories and theorists, made the saga compelling. What was also made abundantly clear was that, although at any given time one of these ideas might be a widely believed orthodoxy, there was no single party line in human geography. This point was hammered home on a weekly basis in my subsequent studies by my historical geography tutor, who would end each lecture with his catchphrase: 'You pays your money, you takes your choice' – meaning that it really wasn't his job to tell us what ideas to believe, but to present us with alternatives, to show us debates. Our job was to think these through, weigh them up, to *make our choice* and then to justify it.[1]

My reminiscence crossed to my other degree subject, geology (and to the other branch of geography, physical geography). Nowhere on the curriculum I followed was there a course on the history and concepts of geology or of physical geography. While tutors would occasionally allude to 'controversies' past and present, from plate tectonics to catastrophic collision as the cause of the dinosaurs' extinction, that sense of doubt and choice was never present (even though, I later found out, at least one of my tutors had a research interest in the history of science). Controversies may come, and at that moment doubt and choice may enter into things, but only for a short while. Order was soon restored, once the 'truth' had been made clear. Some people clung to unorthodox views, but they were generally considered cranks or heretics (see Chapter 6); most practising geologists and physical geographers shared a set of ideas about their subject or profession that went unchallenged. Things that people believed in the past were seen as naïve in comparison with the 'truth' as it is understood in the present. That 'truth' could almost exclusively be taught and learnt without question; a textbook on petrology, or biogeography, didn't have to start by sketching how the ideas it contained had been arrived at. These truths were truly self-evident, uncontested (and uncontestable). Geology and physical geography both consisted of *facts* and *skills* to be learnt – this rock is made up of these minerals, this fossil is this old, the Cambrian period started this many million years ago, this is how to peer down a microscope and see important things. (On the distinction between the teaching of the histories of the natural and social sciences, see also Fuller 1997.)

Now, before anyone gets cross about the gloss I am putting on things, I don't for a minute want to disrespect my physical geography and geology tutors, nor their disciplines as a whole; nor do I want to be seen as claiming that human geography has a higher degree of reflexivity – for that is a large part of what I have been talking about, the extent to which disciplines reflect on their own development and current practice. Both geography and geology are, of course, complex and heterogeneous subjects, branching off into myriad subdisciplines and specialisms. In terms of geography, this splitting – itself an important part of the story of science – begins with that very significant bifurcation: into human and physical geography. I want to take a moment to think about this split, because it has broader significance for understanding science. This dividing is commonplace, and often goes without question. Certainly, geographers are schooled early on in this separation, and as we become more embedded in the discipline we tend to become more closely connected to one or other side of the divide. But what does this split mean, and what's at stake in making it?

Let's begin by looking at what human geography means. *The Dictionary of Human Geography*, a self-styled authoritative resource on such matters, states that it is 'That part of the discipline of geography concerned with the spatial differentiation and organization of human activity and its interrelationships with the physical environment (Johnston *et al.*, 2000: 353). *The Dictionary* moves from this deceptively orderly definition, to sketch the progressive separation of geography into human and physical subdisciplines. The splitting of human from physical geography is a relatively recent phenomenon, and in the UK it only became solidified in the post-war period. In part this relates to the increasing professionalization of geography, especially in universities. It also reflects the broader on-going intellectual separation between the natural sciences and the human sciences – a split which geography ambivalently straddles to this day. On a day-to-day level, this split is manifest in the practices of the two subdisciplines: in the methods and techniques used to do human or physical geography, in the ways it is taught and learnt, written and spoken about.

The two 'halves' of geography have developed along increasingly divergent paths over the past half century, and only recently have calls been made to bring the two closer together. You only have to look at the curriculum of geography as it is taught in schools, colleges and universities, or the vast majority of books and papers that get published, to see the everyday restatement of this separation. I certainly remember being pressured to choose between the two sides at university – and the choice was presented as something with potentially life-changing (or at least career-defining) implications.

Once these two halves of geography had been established as mutually exclusive, the progressive specialization of geography into subdisciplines further reinforced their separateness – so, on the human side we have social or political or transport geography, and on the physical we have biogeography, geomorphology, meteorology. The shorthand way the split is explained (and thereby performed) is to say that physical geography is one of the natural sciences, human geography one of the social sciences – meaning that, depending on the 'route' you choose to take through your studies, geography students can graduate (from UK universities at least) with a BSc or a BA qualification, reflecting their position on the human/physical divide. Despite recent calls to reunite the two tribes of geography (see, for example, Harrison *et al.* 2004), most of the day-to-day business of geography education and employment restates the human/physical, social science/natural science distinction. *You pays your money, you takes your choice.*

If we step back from the particularities of what has happened to geography for a moment – recognizing that geography is a peculiar discipline, located on this fuzzy boundary, schizophrenic about its own identity – we can see that one thing that's being worked through here is what we mean by science, or different kinds of science. The different ways of thinking about physical and human geography, the different stories that are told about them routinely in their teaching and learning, and in their practice, condense a lot of the important themes of this book, as sketched in the introduction. I now want to step away from geography and geology – though given my own training in these disciplines they are ready sources of exemplification, for which I apologize – to look at the idea of science as a particular way of thinking about things.

Science in/and/as society

While there isn't room here to produce a full account of ways of thinking about science, in what follows I will highlight some key issues, ideas and moments. Luckily, there are plenty of useful texts that do a more thorough job, and readers who want more should track them down (see, for example – though be aware of different perspectives – Bucchi 2004; Fuller 1997; Sismondo 2004; Woolgar 1988; Yearley 2005). In my partial account, I will focus mainly on sociological and cultural perspectives, but this does not mean that other ideas, such as those coming from the history and philosophy of science, are not also very useful and interesting. I shall allude to some of these along the way, though here I am at one of my own fuzzy boundaries!

First of all, a simple but important question: what does it mean, to say that science has a sociology, or a history, or a philosophy? To say that science has a history might seem the most uncontroversial of these three, so let's start there. No one would disagree that science has a history, or that the sciences have a set of histories, would they? Science has changed through time, in terms of how it is done, what it says about things, who practises it, how it 'fits' into society, and so on. So, we can track the increasing professionalization and institutionaliza-tion of science, from a largely aristocratic hobby to a multi-million dollar enterprise, over the course of the last couple of centuries. We can note, in fact, the invention of the term 'scientist' as a label to describe someone who does science, in Britain, to the 1830s, notably in the context of a meeting of the British Association for the Advancement of Science (see Bucchi 2004). Now, as historian-philosopher Michel Foucault spent his life's work showing, naming someone as something is a profoundly important step; it is also a key symptom of modernity, of the need for order and classification that modernity felt so acutely (Foucault 1979; see also Chapter 6). So calling someone (or oneself) a scientist is more than simple naming: it calls forth a distinct type of person, and the name becomes the carrier not only of a set of practices (what a scientist *does*) but also a set of characteristics (what a scientist *is like*). As we shall see in Chapter 4, the figure of the scientist has a particular cultural resonance, manifest in popular representation. So this moment of naming is profoundly important in all kinds of ways.

This moment of naming was preceded by the consolidation of a set of practices, of ways of doing science, progressively solidified since the mid-sixteenth century, during the so-called 'Scientific Revolution'. These ways of doing included an emphasis on experiments, on observation, and on communication, often through emerging scientific academies and societies (Bucchi 2004). This solidifying of ways of doing science also led to a progressive bracketing-off of science from other parts of life: science was something only some people did, and it was quite different from doing other things. Here we see an example of a process we shall meet again in this book, 'black boxing' – the shutting off of science (or technology) from society, the obscuring of its workings.

Naming is also linked to organization, as Foucault showed, too (even if his names were given to 'deviants' subject to discipline). Give something or someone a name, and you can begin to sort it or them out; you can classify things or people as having or not having that name – so, these people are scientists because they think this and do this, whereas these people aren't. Once someone can be called a scientist, other people can think about whether they are one too, or want to be one. Scientists can identify with each other, and can help to police the boundaries of science by dismissing some people as pseudo-scientists, cranks or quacks (see Chapter 6). A scientist knows how to perform science, because there are other scientists to emulate; hence doing science can become codified, even ritualized – in rituals such as the experiment, or the public lecture, or the autopsy (to use an example from Foucault (1975) himself).

Of course, it is not enough to name everyone who does science simply a scientist – though at some level that general label does have a lot of meaning. The sciences, and those scientists that practise them, have become progressively subdivided into specialisms. Where once a 'natural philosopher' (one of the things scientists were called before they were called scientists) could dabble in a whole host of different practices, from botany to physics to theology to poetry, science became rigidified into big blocks – physics, biology, geology – which then became subdivided into smaller units (even if, in fact, some of those brought back together those big blocks, as in geophysics or palaeontology, for example). This is partly to do with professionalization (making a career

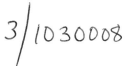

out of following one line of enquiry rather than dabbling), partly to do with knowledge production (it's hard to know everything, easier to specialize), partly to do with modernity (classifying things as precisely as possible, without overlaps and blurry bits), partly to do with institutionalization (the way universities developed and academics laid claim to areas of study) – on all this, see Bucchi (2004). As we have already seen in the case of geography, the subdivision can be problematic, even confusing, especially if you find yourself in a blurry bit.

Nevertheless, as shown, a lot gets invested in maintaining this splitting – not least that, given the logic of modernity, it's believed that the split-up world of sciences we have now is somehow better than the mixed-up blur of the past. This idea, that science gets better as we move through history, is incredibly important, and also something that historians and sociologists of knowledge (including scientific knowledge) have long been battling over. It's one thing that is used to define what Steven Yearley (2005) calls the 'specialness of science' – one of the ways that science is argued to be important, unique even, is that it is continually improving, always working out new things, giving us better facts, better theories. Yet, as Yearley shows, this feature of science proves difficult to support – perhaps especially so today, given the commonly observed and experienced public scepticism about science (a point we shall revisit) – as Yearley (2005: xiii) puts it: 'To say that something is "scientifically proven" is now as likely to be voiced ironically as literally'; nevertheless, he notes, scientists are still seen as being in a unique position – that position hinted at in the opening of this chapter, of being unemcumbered by doubts and competing ideas. Science deals with facts, with truths: 'it is a remarkable achievement for one group in society [scientists] to have created a situation in which that group is believed to speak transparently about how the world is' (Yearley 2005: xii).

Now, to say that science deals in truths or facts, and that the history of science is a tale of improvement through time, implies a number of things. It implies that new discoveries, new observations, new ideas can be added to the existing body of scientific knowledge, progressively 'refining' it. A lot of scientific practice is involved in precisely that set of operations: tinkering with existing knowledge, verifying it, refining it. Thomas Kuhn (1962), a very important figure in understanding science, called this the practice of 'normal science'. The things scientists find out can be slotted in to existing frames, incrementally bettering the explanation a particular frame offers. But there are times when something quite different happens, and Kuhn was especially interested in those times. For quite fundamental scientific ideas – held as truths – do change, and change in a revolutionary rather than iterative way. Truths can be abandoned and new truths installed. Kuhn called this a 'paradigm shift'. Sometimes new evidence makes the existing truth untenable. Now, while people

might skirmish about this, as they have invested an awful lot in believing something as true, oftentimes the outcome is largely one of acceptance of the new evidence and therefore of the new truth; normal science is resumed. Of course, from outside this might look like a strange, even perverse business: a group of people believe something as true, then jettison it and rethink their whole world-view.

As Sergio Sismondo (2004: 14) sketches it, the impact of Kuhn's idea is profound for our view of science: 'Science does not track the truth, but creates different partial views that can be considered to contain truth only by people who hold those views!' We need to pause a moment here, as a version of this insight has been routinely used to bash the supposed relativism of the sociology of science. It's easy to see why this idea might meet with disapproval, even annoyance, from scientists themselves – they don't want to be told (or to appear to be told, for in fact this isn't what Kuhn means) that they are deludedly chasing the truth when in fact it doesn't exist. What it means is we need to recognize that truths are *contingent*, and can be changed, without making them less truthful in the time and place at which they are taken to be true. So, when people believed the Earth to be flat, or the Sun to go round the Earth, these things were as meaningful and true – *as* meaningful and true, not more or less so – than what is commonly believed today about the shape of the Earth or the movement of planets and stars. Trying to clarify the importance of this point became central to a later branch of the sociology of science, known as the **Strong Programme**, which tried to insist that social scientists should treat all kinds of knowledge impartially (though this idea was misrepresented as meaning that all knowledge is equally valid; see Bloor 1991).

The idea that scientific theories and truths can be completely overturned is in itself seen as definitional of science according to another key figure, the philosopher Karl Popper. What makes science 'special' for Popper is exactly this idea, of *falsification*. Unlike other ways of looking at the world, such as Marxism or Freudianism (neither of which he had any time for), science can be proved wrong, and this openness to radical transformation marks science's uniqueness (and superiority). However, as subsequent thinkers worked over Popper's thesis, they noted that falsification can be a tricky business in scientific practice. Normal science, in the Kuhnian sense, is more elastic, more accommodating of anomaly and contradiction, markedly less keen to admit to being proved false. It takes a lot to shift a paradigm, given all that has accreted to it, in terms of life's works, reputations, world-views. In this sense, we can see the idea of paradigms and revolutions in scientific thought as profoundly social. This is not the simple replacement of old ideas with newer, better ones. It involves campaigning, struggles over legitimacy, and brings a whole series of aftershocks, in terms of impacts across and beyond the sciences. As Bucchi

(2004) notes, paradigms may shift for a whole host of 'extra-scientific' reasons, such as the way different scientists publicize their ideas, or the passing of generations. It is also worth noting here that in the Kuhnian model, once a paradigm has been stabilized, the role of scientists as knowledge producers is reconfirmed (Fuller 1997). Moreover, one paradigm is always replaced with another, which must be preferred (ultimately, if not initially) – science must have its paradigms, as 'to reject one paradigm without simultaneously substituting another is to reject science itself' (Kuhn 1962, quoted in Bucchi 2004: 31). I will leave that observation hanging for a while, to give you time to digest its importance.

The science of specialness

Returning to a question posed and then partly answered by Steven Yearley (2005) – the question 'Just what makes science special?' – we can start to think about other ways in which doing science or being a scientist are routinely performed, and how that performance works to suggest first the special status of science as a practice, and second the insulation (or isolation) of that practice from the broader workings and concerns of society. As sociologists of science have been working very hard to show, it is important to think about science as part of society, or as cultural. But this is no easy task, given the specialness accorded to science.

Searching for a way to understand this, Yearley offers a number of possible avenues of enquiry, and I think some of these are worth a look. The first is, by his own admission, an explanation that 'has enduring "everyday" appeal' (2005: 2) – that science is unique because it is founded on *observation* (also known as **empiricism**), thus offering a transparent window on the truth. However, this idea is quickly dispatched by referring to the impossibility of unmediated observation in many branches of contemporary science, such as **nanoscience**; or by the whole problem of ways of looking or seeing; or by examples of conflict between what is observed and what is believed. One set of observations may contradict another, or may counter received theoretical wisdom. While the rebuttal to this problem is either that more observations eventually 'prove' some false and others true, or that repeatedly contradictory observations may provoke a paradigm shift, the underlying critique remains: science does not simply transparently observe then transparently represent whatever is observed. Some things just can't be observed – in theoretical branches of sciences, for example; in fact, all observations are quickly turned into something more abstract, in that they are generalized (as scientists can't observe every instance of the phenomenon under investigation, for ever). A related point concerns

realism: that the things scientists look at and think about are 'real'. This relates to falsification and to the idea of science getting 'truer' as it progresses: falsification is possible because scientists have an imperfect way of thinking about the 'real', and as they get better at that, so they will revise their earlier ideas. The fallibility of science here proves it has unique access to the real world (for realist philosophers this has profound implications for what the world must be like, as the way the world is, is related to the ability of science to know the way the world is; see Yearley 2005 for a more articulate explanation).

The attempt to use empiricism or realism to claim the uniqueness of science draws our attention to a broader issue: how science is practised. Is there a distinct scientific *method*? This was at the heart of Popper's work on falsification, and we've already seen how that played out. Well, then, perhaps science is special because of its workings in a different way: if we accept it as in some ways a social practice, then what social *norms* govern it? How do scientists work as a social group or community? Another key figure in social studies of science had already been thinking about this: the sociologist Robert Merton. In the 1940s, he had identified a moral dimension to scientific practice: scientists worked together in certain ways, with a shared emphasis on how each person's work contributed to the greater good of scientific advancement. These morals, or norms, are really important in terms of how they have been used to make certain claims for science, and also how they have been seen to be undermined by transformations in scientific practice and science's place in society. The four famous norms Merton described are *universalism* (science is impersonal, and so things like 'race' or gender don't impact on the treatment of any scientist's ideas), *communalism* (knowledge is a common good, given freely, not owned by any one scientist), *disinterestedness* (scientists aren't out for glory or riches), and *organized scepticism* (scientists weigh things up, look for proof, take nothing for granted). As Yearley notes, Merton is here describing the professional (rather than scientific) conduct of scientists; more importantly, subsequent studies have overturned Merton's norms, showing some scientists to be, in fact, biased, secretive, money-grabbing or fame-hungry, and so on, while widely implemented notions such as 'intellectual property rights' undermine the ethos of free exchange in science's communalism. Nevertheless, as ideals rather than practised norms, Merton's list is routinely used to cement the specialness of science, as the 'pure' pursuit of equally 'pure' knowledge. As we shall see in Chapter 4, this is turned into a number of negative stereotypes in popular representations of scientists as asocial geeks or crazed megalomaniacs.

It is also important to note here the manner by which sociologists of science have generated their versions of truth; and to think about the norms and practices of a diverse 'field' of study that has been developing and diversifying, particularly in the past half century. At one remove, therefore, we can speak of the 'sociology of the sociology of science'. (We might also add a second category, 'the science of the sociology of science', which looks at the kinds of science that sociologists of science study. For example, Bloor (1991) is unusual in looking at mathematics.) I want to move on now to discuss this idea, and to look at some ways sociologists of science have tried to get us to look at science anew.

Weirding science

Bruno Latour (1987: 97) famously described his favoured method of enquiry as 'follow[ing] scientists around'. While this sounds trite, it is a neat encapsulation of one approach to showing science to be social: empirical studies of 'science in action', to use the title of Latour's book. In fact, there are two variants of this method: one is ethnographic, and involves social scientists entering science's habitats, such as labs, and watching how scientists go about their day-to-day business – what we might call *the everyday life of science* (cf. the science of everyday life, in Chapter 1). The other is to follow their tracks, and is most obviously used in historical studies that examine how past moments of science played out, though it is also evident in studies of present-day science that explore the artefacts of scientific practices, such as documents or machines (of which, more later). Following scientists around has proven a rich if controversial enterprise, yielding a vast body of detailed empirical studies, but also irking some scientists unhappy about being treated as lab rats (one of countless incidents where scientists have complained about sociologists treating them in precisely the way they have long treated their research 'subjects'). Latour's work with Steve Woolgar (1979) playfully announced that they would study scientists as a 'tribe', applying techniques of anthropology to demystify the rituals and practices witnessed in the laboratory.

However, as Sismondo (2004: 87) notes in a useful overview of laboratory studies, 'one of the first things ethnographers found when visiting laboratories was that they, as outsiders, did not see what their informants saw' – there was 'expertise' at work, helping scientists make sense of their own experiments (another rebuff to the idea that science transparently observes 'reality'). A focus on the everyday business, the micropractices (and micropolitics) of lab work also showed how mundane much of it is. It also threw light on the tactics that scientists deploy in their on-going negotiation of experimental work: how do you know if something you haven't tried before is working? This work also showed how scientists rely upon all kinds of objects and artefacts to help them carry out their research, from lab equipment to memos – an insight that was stewed by some sociologists to become part of **actor-network theory** (see

Chapter 3). As Woolgar (1988: 85) writes, the ethnographic approach is at pains to make the familiar strange, using its 'relentless anthropological attention' to look at the mundane goings-on and the 'apparently trivial objects' that scientists make use of. To illustrate, he twice describes a piece of everyday lab equipment, first in the language of the scientist (insider), then as the ethnographer should see it (outsider). It is useful to quote this pair of descriptions in full (with apologies to all those to whom this brings back horrible memories of chemistry lessons, and the unfortunate mishaps of clumsy apparatus usage):

> A pipette is a glass tube with the aid of which a definite volume of liquid can be transferred. With the lower end in the liquid, one sucks the liquid up the tube until it reaches a particular level. Then, by closing the top end with finger or thumb to maintain the vacuum, the tube can be lifted and the measured volume of liquid within it held. Release of the vacuum enables the liquid to be deposited in another beaker, etc.

> Here and there around the laboratory we find glass receptacles, open at both ends, by means of which scientists believe they can capture what they call a 'volume' of the class of substance known as a 'liquid'. Liquids are said to take up the shape of the vessel containing them and are thought to be only slightly compressible. The glass objects, called 'pipettes', are thought to retain the captured 'volume' and to make possible its movement from one part of the laboratory to another.

> (Woolgar 1988: 85)

Looking like an anthropologist upon an unfamiliar scene has proved to be a powerful strategy for 'weirding' science – in the example above, for making us look afresh at the operation of a pipette, something carried out routinely in countless labs apparently without a second thought. Aside from the humour and beauty of describing mundane actions and artefacts in this way, laboratory studies have produced some very significant observations about the practice of science, including that science is *constructive*, rather than merely descriptive (Knorr-Cetina 1981). It does more than passively observe and transparently describe the world; its actions 'are actively engaged in formulating or constructing the character of that world' (Woolgar 1988: 87). Moreover, as Knorr-Cetina says, labs are *materializations* of past science, in terms of the development and refinement of equipment, techniques and so on. That a glass pipette reliably holds a certain measurable volume of liquid until released is the result of experiment, manufacture and use: if a scientist didn't 'trust' the pipette to hold the liquid, or its gradations to accurately reflect the volume of liquid held, then the simple procedure twice described above could not take place. A single operation within a single experiment in a single laboratory, therefore, is the outcome of multiple prior processes, decisions and selections.

An emphasis on anthropological method also encouraged some researchers to see science as a culture, and therefore to deploy modes of cultural analysis to sites like laboratories (e.g. Traweek 1988; see also Reid and Traweek 2000), and to analyse the 'cultural embeddedness and the cultural circulation' of science (Balsamo 1998: 291). This includes, among many other things, an emphasis on discourse and representation – on how science is talked about and written about, how it is depicted textually, visually, orally and so on, both by scientists and by various non-scientists (on the latter, see Chapter 4). Studies have shown, for example, how scientific 'facts' acquire a veneer of truth through repetition, and through the progressive casting-off of uncertainty and contingency: something that took an entire journal article to carefully, minutely explain may be reduced to a sentence in a later textbook and a sound bite on TV news (there are plenty of examples of that in this chapter!). The tidying-up that moves half-thought-through ideas to clear, unequivocal arguments (which can be seen as a move from 'private' to 'public' discourse; see Bucchi 2004) is exactly the kind of move that can be tracked through 'following scientists around'.

As Woolgar showed above, issues of rhetoric and representation extend to the writing-up of the sociology of science, too. In fact, this is another key way that science can be 'weirded' – by writing about it otherwise (see Clough 2001 for a critical discussion). So, in the example from Woolgar, the pipette, and ideas like volume and liquid, are explained as if you'd never heard of them before. The description could go even further, and assume you don't know what glass is, or how a vacuum works. It might even try to explain why this process is repeated in a seemingly ritualistic way through the laboratory day, in the way that anthropologists have tried to understand other 'tribal' rituals. Such techniques can work to reveal, to borrow the title of a book, 'naked science' (Nader 1996). Steve Fuller (1997) does just that when he writes mischievously about uncovering a paper written by Martian anthropologists trying to make extraterrestrial sense of Earth science (with a Mork-ish kind of puzzlement, for those of you old enough to remember *Mork and Mindy*). This often hilarious but also deadly serious writing experiment compares science with religion, and comments on both the public understanding of science and on 'popular science'. Fuller describes the paper rather than mimicking Martian anthropology, thereby retranslating its findings back into earthly contexts.

A similar tactic is attempted in Stephen Hilgartner's (2000) *Science on Stage*, which rewrites US controversies on diet and health, and the role of science advisors in such debates, scripting the events as a drama, revealing the tense stand-offs between, among others, nutritionists and food producers. In explicitly foregrounding ideas of script, props, performance, audience, backstage and so on, Hilgartner's dramaturgical account produces a different take on Fuller's Martian paper – but with the same focus on how science has come to assume and maintain the position of authority it has in society. As Hilgartner argues:

> The notion of performance offers an apt metaphor for analyzing the production of science advice. To win confidence in their audiences, advisory bodies actively display their competence and credibility. They present themselves as knowledgeable and trustworthy. They comport themselves in ways that exhibit their integrity and good judgement ... Self-presentation is central to the science advisor's work.
>
> (Hilgartner 2000: 8)

Hilgarnter's study is also interesting because it moves away from the laboratory as the site for the performance of science, instead focusing on a number of places where science advisors meet their 'audiences'. In exploring the interface between science and other social actors (government, press, public), *Science on Stage* adds to the studies of the more closed-off world of the lab.

This is in line with Sismondo's (2004) comment on the move of ethnographies of science away from the lab, into the field (among the examples he cites is Goodwin's (1995) work on an oceanographic research vessel). Sciences based 'in the field' therefore provide new opportunities for following scientists around, new opportunities to watch how scientists assemble their ideas. Thomas Raab and Robert Frodeman (2002) provide an evocative (for me, at least) example of this in their application of **phenomenology** to the discussion of fieldwork in geology. Watching geologists trying to make sense of fragmentary evidence (rock samples, landscape measurements) in the field, brings to the surface the use of embodied skills and of seemingly non-scientific things like intuition. Geology depends on inference and analogy, since it is often engaged in working out (as in **plate tectonics**) what happened in the long-distant past through looking at 'clues' in the present. For example, as Raab and Frodeman put it:

> In field work, experimentation in a strict sense is not possible: geology confronts a unique set of events (i.e., what happened at a given location) in the distant past ... In conceptualizing the huge time spans and space dimensions active in geology (compared to a human life span), one has to rely on analogies to one's own everyday experience. Geologists manipulate blocks in their imagination in order to understand plate tectonics; and in having to make sense of geological time, they rely upon analogies such as comparing the earth's age to the periods of the day or year. Without such analogies practically no historical geological interpretation would be possible.
>
> (Raab and Frodeman 2002: 71)

Perhaps this is partly why, as Mitchell (1998: 207) observes, palaeontology 'is a highly sensitive barometer of technological and cultural change' – because those are the realms from which geologists draw imaginative analogies to convert fragments of bone and shell into living, moving creatures.

In stressing the embodied knowledge that geologists take into the field, Raab and Frodeman show that practising Earth science involves the 'feel' of things, of sensing the 'lay of the land': 'Fundamental [to field geology] is the mere act of walking the Earth: seeing the land and outcrop from moving and shifting perspectives where a variety of scales intersect and play off one another' (76).[2] A lot of this is imaginative and experiential work – geological knowledge is practised by embodied geologists in the field, who 'read' the landscape in order to understand past events that they can never know directly. Other sociologists of science have referred to this as tacit knowledge, such as experience, which is deployed alongside other forms of knowledge as scientists work in the lab or in the field (see Sismondo 2004). As with laboratory studies that emphasize the subjective skills of observation, measurement and inference, then, field work studies give us a rich depiction of what Raab and Frodeman call the 'intrinsic fuzziness' (79) of much science.

An important question that is catalysed by this discussion of tacit and embodied knowledge concerns how those knowledges are accrued, in what contexts, and by whom. This brings us back to the figure of the scientist, the embodiment of scientific knowledge, the performer of scientific practice. Can *anyone* be a scientist, or at least a potential scientist, or is there something 'special' about some people that makes them, in Steve Fuller's (1997: 34) tart words, 'decent little physicists'? If we think about how this question has been translated through education, or through the idea of 'popular science' (see Chapter 6), we at once see that there is an idea that everyone could be a scientist, but at the same time an idea that scientists are somehow different, special. Fuller notes in this regard that being 'good at science' has long been taken as a measure of general intelligence in school-level education in Britain. Of course, there are strikingly similar (but at the same time totally different) arguments about, among other things, whether everyone could be an artist, or whether artists are special people (on the comparison between the scientist and the artist, see Jones 1998). One way in which societies like ours have managed this question is through making a distinction between amateurs and professionals – the hobbyist painter or weekend electronics tinkerer, versus the art school graduate or lab worker (see Chapter 6).

Turning science – or art, for that matter – into something to be studied, and then into a profession, involves a number of significant manoeuvres. Fuller (1997) notes, for example, the progressive professionalization of science, the marginalizing of amateurs and hobbyists not affiliated to academic institutions or professional bodies, and also the irony of this move in light of the past emphasis on the necessity of leisure to proper scientific contemplation, not to mention the long history of tinkering by amateurs that has contributed innumerable scientific discoveries through the years. As he explains it, the crunch time in Britain was the nineteenth century, during which two 'convergent tendencies' exerted pressures on scientific enquiry – 'an industrial "push" and an academic "pull" ' (37). The industrial revolution brought new uses for science, in terms of improving productivity and so on; universities effectively colonized science training and education, making themselves the proper (and only) place that could produce scientists, and in terms of industry produce future potential inventors, by explaining the underlying laws and principles of nature that industrialists sought to exploit or manipulate. Such a pincer-movement left little room for hobbyists and amateurs, although new spaces were opened up for 'non-traditional' ways of learning and doing science throughout the last century. Popular science magazines and clubs, for example, helped remake amateur science as a leisure pursuit (in the guise of 'rational recreation') and as a space for consumption (Fahey *et al*. 2005). The crucial point here, however, is that the definition of amateur and professional has been *hardened*. Another important distinction was also effected by the move from amateur to academic science: the amateurs of the period 1600–1800 were nonspecialists, 'natural philosophers' who could dabble in any branches of science; one consequence of the institutionalization and professionalization of science was its compartmentalization. A university physicist would not be expected to be an expert in biology, and moving from one specialism to another was rare (Woolgar 1988). (Of course, one of the paradoxes of science today is its hybrid nature, with 'dedifferentiation' as well as differentiation – artificial intelligence, for example, combines biology, psychology and computing, while cosmology reaches even further, into philosophy; see Crook *et al*. 1992.)

Woolgar (1988) determines three phases in the history of the social organization of science: the amateur phase (1600–1800), the academic phase (1800–1940) and the professional phase (since 1940). This latter phase marks decisive shifts in things like the funding of science, the idea of a split between 'pure' and 'applied' research, the rise of corporate research and development (R&D), military and industrial financing, applications and involvement, the changing relation between science and the state (and the rise of science policy), the mixing of science and enterprise, the idea of intellectual property, and so on – a phase widely seen as culminating in vast, expensive projects known as **big science**. This 'post-academic' science is often at odds with the old Mertonian codes of scientific conduct, and some commentators argue that the face of science is irrevocably changed in this new era of 'postmodern technoscience' and maybe even 'postscience' (Crook *et al*. 1992; Sassower 1995). As Best and

Kellner (2001: 113) write, in their review of 'postmodern science', 'some forms of "science" [today] are no longer "scientific" according to the classical norms' – as theoretical physics drifts towards metaphysics or theology, for example. The purported postmodernizing of science must, it seems, be our next port of call.

Postmodern Science Wars

Postmodernity, postmodern, postmodernization: in order to understand how contemporary science works, we need to understand a little of these complex and vexed terms. An entire academic industry has spent the past few decades debating these terms, their meanings and implications, perhaps most notably in the humanities disciplines. And still they argue about the fine points and the big picture. Careers and reputations have been made – or at least enhanced – by showing us the history of the idea of the postmodern (Bertens 1995; see also Sim 2002), by pointing to its discontents (Bauman 1997), even by being dead set against it (Callinicos 1989). In a characteristically postmodern move, these different positions can all be accommodated, along with countless others, as postmodern*isms* proliferate (Sassower 1995). But postmodern science? How can the characteristics of postmodernity – uncertainty, irony, pastiche, irrationality – be docked with scientific knowledge and practice?

For some critics, they can't. Science is unamenable to postmodernizing; it is the supremely modern enterprise. It is about truth and stuff like that; stuff that the postmodern has no truck with. But other commentators argue that we live in postmodern times, and to think that science is somehow insulated from postmodernization is a delusion. The task becomes instead to *understand* postmodern science, not simply dismiss it as an oxymoron. Luckily, we have some guides to help us through this. The first, and among the most famous (or infamous, depending on how you see it) is Jean-Francois Lyotard, author of *The Postmodern Condition: a Report on Knowledge* (published in French in 1979 and in English in 1984), one of the cornerstone statements about postmodern science, with its often-quoted aphoristic definition of postmodernity as 'incredulity towards metanarratives' (Lyotard 1984: xxiv). Metanarratives are the Enlightenment's transcendent, universal truths, truths that underpin Western civilization – things like reason and progress (and indeed the idea of 'truth' itself). In trying to understand the status of knowledge in an age where people no longer believe in things like progress (having witnessed things getting worse as well as better), Lyotard focuses on the idea of legitimation: how do some forms of knowledge get made more believable than others? He draws on the philosopher Ludwig Wittgenstein to show that knowledge is produced through 'language games' – how things are made 'real' by being spoken about in certain ways. He focuses on scientific knowledge, arguing that this too is legitimated by narrativization, by language games: 'Scientific knowledge cannot know and make known that it is the true knowledge without resorting to the other, narrative, kind of knowledge, which from its point of view is no knowledge at all' (Lyotard 1984: 29). So part of the language game science is playing is pretending it is *not playing*; that playing such a game is beneath it.

Incredulity towards transcendent metanarratives makes it harder for science to legitimize itself, to justify its role and position in society as producer and custodian of truth, because the workings of science (observation, reason, progress, truth) belong to that category of metanarratives that we are incredulous towards! Modern science still attempts to assert its legitimacy by appeals to those metanarratives, but such appeals fall largely on deaf ears, and modern science is reduced to what Lyotard refers to as performativity, by which he means a narrow instrumentalism ('applied' science at the service of capitalism, for example) with no room for what is called 'blue skies research' – research without a clear goal. But all is not lost, for postmodernity offers amazing opportunities for a new science, a postmodern science, and a lot of its sky is indeed blue. Lyotard powerfully and suggestively sketches this new science:

> Postmodern science – by concerning itself with such things as undecidables, the limits of precise control, conflicts characterized by incomplete information, 'fracta', catastrophes, and pragmatic paradoxes – is theorizing its own evolution as discontinuous, catastrophic, nonrectifiable, and paradoxical. It is changing the meaning of the word *knowledge*, while expressing how such a change can take place. It is producing not the known, but the unknown. And it suggests a model of legitimation that has nothing to do with maximized performance, but has as its basis difference understood as paralogy.
>
> (Lyotard 1984: 60)

In this heady formulation, then, postmodern science isn't afraid of unknowns and paradoxes; it doesn't seek straightforward truths, but recognizes, even embraces contingency, dissensus and anomaly; and it is not merely instrumental, about improving productivity or whatever, but is based around 'paralogy' – stimulating conversation out of which new ideas and ways of thinking emerge (Shawyer n.d.).[3]

Stephen Crook *et al.* (1992: 206) toe a similar line in their discussion of the *disorganization* of science, for example in terms of the blurring of disciplinary boundaries and, crucially, in the way that 'public support for science and science funding becomes both more volatile and more instrumental'. This is an important dimension to science's 'legitimation crisis': it is no longer taken as read that science is a good thing, worth supporting and funding. Scientists have to work harder and harder to convince their 'publics' of the value (increasingly rewritten as usefulness) of science, and moreover that its outcomes will produce benefits – and the benefit of better scientific knowledge is no longer sufficient. Equally importantly, it must be seen to be incapable of producing disbenefits, whether these are intentional or accidental. As Raphael Sassower (1995) shows in his book on attempts to secure funding and support for a US superconducting supercollider – a huge science facility concerned with exploring the building blocks of matter – there's a lot of public relations work and political glad-handing now required, and a lot of what he calls 'translation' (language games), in order for scientists to legitimate spending billions of dollars accelerating subatomic particles on collision courses so they can capture that fraction of a fraction of a second when those particles smash and break up into even tinier bits. Perhaps the closeness in lay terms of this project and the splitting of the atom – which led to the Bomb – makes the enrolment of popular support and the obscuring of potential disbenefits particularly difficult in this case (see Chapter 5).

Moreover, the 'anything goes' tone of Lyotard's vision of postmodern science must not be read in that simplistic way: while some branches of science have become disorganized and postmodernized – chaos theory, complexity theory or quantum mechanics being the usual touchstones in that regard, though not uncontroversially (Sim 2002; Yearley 2005) – there's still an awful lot of rational thought at work in scientific practice, still a lot of boundary policing between science and pseudoscience, still a quest for truths (see Chapter 6). Not all scientists feel the legitimation crisis in their day-to-day lab lives, though they may feel it when applying for a government grant, or talking to a non-scientist neighbour, for example. It should perhaps be added that not all those engaged in social and cultural studies of science subscribe to the Lyotardian line, either (see, for example, Yearley 2005).

The divergence of stances on the current role and status of science and scientific knowledge, and also of social and cultural studies of science, came to a head in the USA (and elsewhere) in the mid-1990s, in a conflagration known variously as the Science Wars or the Sokal Hoax. At one level this marks merely the latest in a long line of spats between science and varieties of science studies (Fuller 2000). This line stretches back through C.P. Snow's (1963) famous description of the incompatible 'two cultures' of scientists and humanists, and arguably right back to much earlier discussions about the role of science in society, whether staged via fictions like *Frankenstein* or debated by philosophers and theologians. But since it occurred in postmodern times, the Science Wars in their Sokalist manifestation offer a test-bed to revisit Lyotard's ideas, and also to revisit some of our fundamental questions, such as: *what's at stake in thinking about science as culture?*

There has been an immense amount of commentary on the Science Wars and the Sokal Hoax, academic and popular, scientific and non-scientific. If I retell the tale here with brevity, readers wanting to know more can find good summaries and discussions in Ross (1996) – a revised version of ground zero of the Science Wars – Segerstråle (2000) and the little pocketbook by Sardar (2000), as well as countless other books and articles. The tale has multiple beginnings. One was the publication of a book, *Higher Superstition: the Academic Left and its Quarrels with Science* (Gross and Levitt 1994), which rounded on 'science bashers' and continued a mode of argument about the dangers of 'irrationalism' and antiscience or pseudoscience in popular culture and in the academic humanities today. There was also controversy surrounding an exhibition at the Smithsonian, *Science in American Life*, which opened that same year (Gieryn 1998).

Then there was a special issue of a key US journal of the 'academic left', *Social Text*, two years later. The theme of the issue was the Science Wars, partly in response to Gross and Levitt. Among the papers received by the editors was an unsolicited manuscript by Alan Sokal, 'Transgressing the boundaries: toward a transformative hermeneutics of quantum gravity' (Sokal 1996a), which was duly published in the Science Wars issue alongside essays by well-known figures in science studies such as Steve Fuller, Emily Martin and Sharon Traweek. Then came the punchline: Sokal published a second article, 'A physicist experiments with cultural studies' (Sokal 1996b) in the magazine *Lingua Franca*, in which he exposed his own hoaxing – the article *Social Text* had printed was hokum, a bricolage of inaccurate science stitched to quotes picked randomly from key cultural theorists and postmodernists (it has an enormous bibliography!). The revelation caught a fair amount of media coverage, and caused a lot of people a lot of bother. Sokal's point was an old one: leave science to the scientists. Here is his justification for perpetrating the prank, in the introduction to a later book co-written with Jean Bricmont in which they attack various theorists for 'misusing' science:

> For some years, we have been surprised and distressed by the intellectual trends in certain parts of American academia. Vast sectors of the humanities and social sciences seem to have adopted a philosophy that we shall call, for want of a better term, 'postmodernism' . . . To respond to this phenomenon, one of us (Sokal) decided to try an unorthodox (and admittedly uncontrolled) experiment: submit to a fashionable American cultural-studies journal, *Social Text*, a parody of the type of work that has proliferated in recent years, to see if they would publish it. . . . The parody

was constructed around quotations from eminent French and American intellectuals about the alleged philosophical and social implications of mathematics and the natural sciences. The passages may be absurd or meaningless, but they are nonetheless authentic.

(Sokal and Bricmont 1997: 1–3)

The heat generated by the hoax and its revelation focused a longer-running attack on the academic left in the USA (with previous battles over political correctness, for example) down on to its complicity in undermining the legitimacy of science precisely through studying it through a social or cultural lens. Such a position takes as read the notion that social and cultural studies of science are antagonistic to science, awash with anything-goes relativism, concerned with exposing science as trickery, as language games.

In a sense, Sokal was right, but he was also dead wrong: as the quote from W.J.T. Mitchell at the top of this chapter says, the social and cultural study of science is not about discrediting science, it is about *understanding* it. Exposing cultural critics as fraudsters on the grounds that they don't fully grasp the complexities of scientific theories is a straightforward act of boundary-policing: keep off! (Ironically, of course, this doesn't stop Sokal trampling all over cultural theory, to question its legitimacy.) Ziauddin Sardar (2000) rightly reads the Science Wars as proof of the erosion of science's credibility and legitimacy, its move into a 'post-normal' phase akin to the postmodern science described by Lyotard earlier. Other critics are more sanguine, noting for example that science studies has had very little effect on how science is practised; 'We get the blame without getting results' is how N. Katherine Hayles (1996: 235) sums it up. For science studies exists primarily as a fringe topic even in sociology or cultural studies, appearing only patchily on undergraduate curricula or in core textbooks of the disciplines. It is all but invisible in science, except during brief firestorms like that fanned by Sokal. However, that such a marginal academic specialism could make the front page of *The New York Times* or *Le Monde*, as the Sokal Hoax did, tells us that there's more at stake here. What's at stake is not rubbishing science, and nor should it be rubbishing science studies; what's at stake is *including* science, making it more apparently what it so plainly is: a human, which is to say social and cultural, enterprise. This in no way diminishes the centrality of science as a site of knowledge production; it only asks that we see it as precisely that: *one* site where knowledge is *produced*, by people (and artefacts).

Give peace a chance

That science studies, and the Science Wars, have had little or no impact on the practices of science should give us pause for reflection. George Levine, one of the contributors to the *Social Text* special issue, asked the pertinent question in the title of his essay: 'What is science studies for and who cares?' (Levine 1996). Identifying science as a 'concrete social practice', in Mitchell's words, means we must also recognize it as other things, too: as political, for example. Steve Best and Douglas Kellner (2001: 17) round this out nicely, arguing that science studies should 'analyze science and technology from the optic of its impact on politics, identities and everyday life', to which we must add the corollary: it should also analyse the impact of politics, identities and everyday life on science and technology. Implicit in cultural studies of science (and technology) – and more often than not *explicit*, too – is a commitment to such calls. Contrary to Sokal's boundary policing, as Best and Kellner put it, 'science and technology are too important to be left to scientists and technocrats themselves' (18). And Ann Balsamo (1998: 294) writes that seeing science as culture is about acknowledging 'that science, technology and medicine represent the central institutionalized sites of ideological work in contemporary culture'. That tells us succinctly what science studies is for, and the answer to the question 'Who cares?' must surely be 'All of us'.

Further reading

Best, S. and Kellner, D. (2001) *The Postmodern Adventure: Science, Technology, and Cultural Studies at the Third Millennium*. London: Routledge.

Bucchi, M. (2004) *Science in Society: An Introduction to Social Studies of Science*. London: Routledge.

Fuller, S. (1997) *Science*. Buckingham: Open University Press.

Raab, T. and Frodeman, R. (2002) What is it like to be a geologist? A phenomenology of geology and its epistemological implications, *Philosophy and Geography*, 5(1): 69–81.

Reid, R. and Traweek, S. (eds) (2000) *Doing Science + Culture: How Cultural and Interdisciplinary Studies are Changing the Way We Look at Science and Medicine*. London: Routledge.

Ross, A. (ed.) (1996) *Science Wars*. Durham NC: Duke University Press.

Sardar, Z. (2000) *Thomas Kuhn and the Science Wars*. Cambridge: Icon.

Sassower, R. (1995) *Cultural Collisions: Postmodern Technoscience*. London: Routledge.

Segerstråle, U. (ed.) (2000) *Beyond the Science Wars: The Missing Discourse about Science and Society*. New York: SUNY Press.

Sismondo, S. (2004) *An Introduction to Science and Technology Studies*. Oxford: Blackwell.

Woolgar, S. (1988) *Science: The Very Idea*. London: Routledge.

Yearley, S. (2005) *Making Sense of Science: Understanding the Social Study of Science*. London: Sage.

Notes

1 That tutor was the great Tony Phillips, authority on Victorian field drainage systems. Funnily enough, I mistakenly signed up for his course in historical geography thinking it was about the history of geography, though in the end that didn't matter, for the course was amazing, and started my apprenticeship into academia proper.

2 While I really enjoyed this article, it also reminded me of the time I spent three weeks in the rain-sodden Lake District attempting and failing to marshal my very limited embodied geological knowledge to interpret and map a patch of the Earth. My experience of walking and 'reading' from the landscape was primarily one of exhaustion and utter bewilderment, as the outcrops and samples refused to render up their stories to me.

3 Here is Shawyer's (n.d., n.p.) elegant definition: ' "paralogy" means a flood of good ideas that are inspired by conversation. Postmoderns, he [Lyotard] tells us, have a quest for "paralogy", a hunger for stimulating conversation and ideas that work in a satisfying way. To get those ideas paralogists often share an irreverent attitude towards well accepted theories, breaking them up and recombining them in revolutionary new ways. The point of paralogy is to help us shake ourselves loose of stultifying traditional frameworks that we have come to take for granted in order to enhance our spontaneous creativity.'

3 | THINKING ABOUT TECHNOLOGY AND CULTURE

We live our lives in a world of things that people have made. As human beings, we have both relations to each other (the set of relations we call 'society') and also relations to the things we have made and to our knowledge of those things – in other words, relations to 'technology'. Almost everything human beings do – paid work, domestic work, warfare, childcare, healthcare, education, transport, entertainment, even sex and reproduction – involves relations both to other people and to things.

(Donald Mackenzie and Judy Wajcman)

A fridge is being delivered

A new fridge is being delivered here today; and not just any fridge – a stainless steel, double-doored, 'American style' fridge-freezer. It has a thing on the front that dispenses cold water and – get this – *cubed and crushed* ice. As I sit writing, I keep glancing from screen to window and then to wristwatch, wondering when the delivery will come. When I woke up this morning and remembered that today was new fridge delivery day, I began to also ponder the twin lenses of this chapter: technology and culture. I thought first about the countless technologies and technological systems implicated in today's anticipated events. Of course, there's the fridge itself, and the technologies of domestic refrigeration, plus its offspring in terms of the production of refrigerated and frozen foods, and so on (see Cowan 1985; Shove and Southerton 2000). Then there's the lorry that will bring the fridge, with its tail lift that enables the delivery people to get such a heavy object down and out. There's the telephone – I am eagerly awaiting a call to tell me what time they'll be here, albeit vaguely. The fridge was bought over the Internet, found using a search engine, paid for using 'secure' online systems. Behind these more-or-less immediate technologies, there's all the assorted systems and infrastructures, the phone lines and broadband networks, traffic management systems, shipping and storage facilities, logistics, credit checking databases, etc. (On complex technological systems, see Mackenzie and Wajcman 1999; on the importance of studying technological infrastructure, see Star 1999.)

What, then, can we say about the fridge from a cultural angle? Let's start with *this* fridge: you can see, hopefully, even by the brief description above, that the fridge has long transcended the purely utilitarian functionality common to so-called 'white goods' until comparatively recently. The fridge, in common with many domestic appliances, has been made over, *lifestyled* (Bell and Hollows 2005). Its ability to do its job – keeping things cold, or frozen – is taken for granted, and having a fridge and a freezer at home has become 'normal' (Shove and Southerton 2000). What differentiates fridges is their look, aspects of their performance (especially their eco-friendliness and efficiency), and the add-ons and peripherals (in this case, that includes the water and ice dispenser). The stainless steel outside is crucial to its appeal, given the current fetish for industrial-looking appliances (where stainless steel signifies industrialness in kitchens and their appliances). Note also that the fridge is described as 'American-style' – and not just by me, but by the sales team and marketers who made it seem so desirable. Do fridges have nationalities? Recent work on national identity and popular culture, such as that by Tim Edensor (2002), has explored just this: in his case, he looks at the nationalities of cars, like the British Mini or the German Volkswagen. In the case of this fridge, American-ness signifies particular things, positive things. (We Brits have an enduring love/ hate relationship with America and its cultural products, spanning at least the past 50 years – see Webster 1988.) The fridge is also *big* – itself part of its Americanness, like the bigness of American cars – which is itself an interesting twist on one dominant trope of technological innovation, which moves towards progressive miniaturization. In fact, it seems almost out of scale, perhaps unintentionally marking the Englishness of the rest of the kitchen by making it all seem so small (on the strange effects of scale, see Self 1995). It stands out and proud, bucking the trend for disappearing appliances in built-in kitchens (Shove and Southerton 2000). No wonder they are given the aggrandizing name 'food centers' in America.

Thinking about fridge cultures, of course, also means attending to use, to the domestic, everyday life of and with fridges. The domestic refrigerator and the deep freeze have radically transformed practices of food shopping, storage, cooking and eating. Living without a fridge makes this very clear. The fridge and freezer make possible the familiar 'big shop', and so encourage the economies of scale of the hypermarket (Humphery 1998). Put a different way, they free us from the burden of having to shop daily for perishables. Or they have contributed to the dominance of off-centre retail 'sheds' and the demise of local shops. Fridges allow a more varied diet, since freezing preserves foodstuffs.

They bring convenience to household 'food work'. Or again, from a different angle, fridges have contributed to the erasing of seasonality from cooking and eating, disconnecting us from the Earth's natural rhythms and instead reflecting the frenetic rhythms of 'time-poor' households (Shove and Southerton 2000). And, of course, there's the fridge as an emblem of bad technology, of CFCs and the growing 'fridge mountains' awaiting 'safe' disposal – growing even quicker now the fridge has been lifestyled, so its obsolescence is as likely to be stylistic as functional. The fridge should be green, too.

In their book *Tools for Cultural Studies*, Thwaites *et al.* (1994) also pondered the domestic fridge, noting that it performs many other functions in the home – functions not 'written in' to its functionalist design, but nonetheless common-place. It is so beautiful a description, I want to quote it at length here:

> The top of a refrigerator nearly always comes to serve as a shelf. . . . [B]ecause the area on top of a fridge is 'not really' a storage area as such, it can serve as a miscellaneous catchment area for things which don't have any other easily assignable place, either in the kitchen or elsewhere. This surface is often a metre and a half or more above the ground, so it also serves as the ideal place for items which need to be kept out of the reach of children, household pets or insects.
>
> (Thwaites *et al.* 1994: 184)

Let's pause here and look at what's been said so far: the fridge has an important unofficial second function, as a high-up shelf for heterogeneous oddments, or for things that need protecting from certain people or things (though, it should be noted, height offers no protection from winged insects and pets!). Immediately we see how pieces of domestic technology lead what David Noble (1984) calls a 'double life'. On the one hand they do the thing they're supposed to: keeping food cold or frozen[1] – even though the reasons for 'needing' this function have changed over time, from freezing gluts of home-grown produce to provisioning ready meals (Shove and Southerton 2000). But they also do other, unintended things, as we'll see later. Now back to the quote:

> But it's the white enameled vertical surfaces of the refrigerator which most clearly show improvisation at work. Like a caddis-fly larva, a fridge gradually covers most of its visible surface with bits and pieces gathered from its environment: notes, magnets, information, decoration. The fridge becomes a clearing-centre for household information . . . It is [also] a sort of semi-public scrapbook of personal bric-a-brac, open to and even inviting the visitor's gaze . . . Brightly coloured magnets even mean that the refrigerator door can be a temporary child-minder under the super-vision of a parent preparing food – not to mention a walking aid for the

very young. And, of course, the fridge is always one of the focal points at
parties.

(Ibid.: 185)

I *love* this passage – the fridge as caddis-fly larva, accreting detritus on to itself,
and those last, specific uses of the fridge, for kids and at parties. But, track back
to the start of this part of the extract, and its description of 'the white enameled
vertical surface' of the fridge door. Such an anonymous, blank surface invites
such accretion, such bricolage. But what about my new, shiny, stainless steel
fridge? I start worrying about fridge magnets (themselves an amazing little idea,
now endlessly proliferating and solving many a last-minute holiday memento
purchasing panic). Can fridge magnets be permitted on the new fridge? I somehow doubt it, at least while it is still new. So the lifestyling of fridges, which has
so often involved adding 'designer' features to their front faces, has transformed
the everyday double life of the fridge as a message centre and scrapbook. Will
we turn to using some other white goods instead: washing machine magnets,
anyone? (Word of warning: someone told me you should NEVER put fridge
magnets on a microwave.) So, as well as anxiety about the possible fridge
magnet policing problem, I'm also left wondering if household life can in fact
continue without such an ad hoc message board. Where will we put all those
things?

Thwaites *et al.* provide a nice audit of one fridge's unintended matter, listing
on the front and exposed side of the fridge the following:

[On the side:] magnetic cards from local pharmacist, milk vendor,
2 plumbers, ice-cream vendor, hamburger chain; takeaway menus from
local pizza and seafood places; memo board with black felt pen on string;
two school photographs, eight non-advertising magnets (letters, numerals,
animals, etc.). [And on the door:] magnetic cards from gas utility, ice-
cream vendor (again), hamburger chain (again) and local seafood takeaway; magnetic calendar from local Lebanese takeaway; magnetic frame
with photograph of children in bath; three pictures by children; price list
from school tuckshop; sticker from cereal packet (non-removable); sixteen
non-advertising magnets.

(Ibid.: 186)

Setting aside the horror that the words 'non-removable' spark in the newly-
fridge-proud, think about how much of this audited household's day-to-day
running is **delegated** (an important word we'll return to) to the fridge. Also note
the already remarked upon semi-public nature of this montage – displaying
pictures and photos as well as perhaps saying something about this household's
takeaway consumption (as well as its membership). And all this before you've

even opened up the fridge and read off yet more things by examining its contents.

I have taken this strange and rambling start in order to lay out, implicitly, the key themes of this chapter. As in Chapter 2, my concern here is to explore what's at stake in thinking about technology as culture, and then to show you some ways that this thinking could follow. So, it's time to move away from the fridge, for now at least, and ask that most basic of questions . . .

What is technology?

In an essay with that very question as its title, Stephen Kline (2003: 210) opens up the problem of definition, noting that the term 'technology' can be variously used to describe 'things, actions, processes, methods and systems'. Perhaps most commonly (and common-sensically) it is used to refer to artefacts or objects, or what he names hardware. We routinely use the word to refer to *things*, most often things that are 'high tech', that is, those things whose 'technologicalness' is foregrounded or emphasized: gadgets and gizmos, digital things, shiny new devices (Lehtonen 2003; Michael 2003). Of course, there are many other things around us that are equally technological, but whose 'technologicalness' is backgrounded, or taken for granted. As Timothy Taylor (2001: 6) puts it, 'one of the ways technology works in Western culture is to call attention to itself when new, for at that moment it has no social life . . . After a period of use, most technological artifacts are normalized into everyday life and no longer seen as "technological" at all'. The link between technology and newness is so strong, he adds, that

> if today someone says that she is interested in technology, this is generally taken to mean computers or other kinds of current digital technologies. But if she says she is interested in, say, kerosene lamps, then most listeners would assume she is interested in antiques, not technology, even though, of course, kerosene lamps were once considered cutting-edge technology.
>
> (Taylor 2001: 7)

That older technologies are rendered non-technological is in part to do with the narrative of technological improvement, in which things get better by being more technological, and also to do with the comment Taylor makes about their 'social life' – as technological artefacts become embedded in their users' everyday lives, so their 'technologicalness' fades. (See Michael 2003 for a different take on 'exotic' and 'mundane' technologies and their relationship to time.) At the same time, however, large parts of that very 'technologicalness' are kept hidden from users, through a process called black boxing.

In science and technology studies, black boxing is borrowed from engineering, where a black box is a simple, predictable, input-output device; something, as Sergio Sismondo (2004: 97) puts it, 'the inner workings of which need not be known for it to be used'. Technological artefacts are commonly black boxed, in that 'ordinary users' know very little, if anything, about how they really work. Think about the everyday technologies you use – the mobile phone, the PC, the iPod. Can you explain how they work? Black boxing occurs in part because we don't need to know. We know the inputs and expected outputs, so the machines have 'usability'. There are times when we might potentially need to know more, such as when they get broken or malfunction, but there is an army of technicians on hand, whose services and expertise we can buy in, not to mention techno-savvy friends who we call on as 'warm experts' (Lehtonen 2003). But black boxing is about more than this; as Sismondo writes, it's also about inevitability – it's about saying that phones or iPods or fridges look and work as they do because they offer the only solution to a set of problems – how to communicate using your voice in real time while out and about; how to store and retrieve pre-recorded music in a digital format portably; how to keep foods cold or frozen. To 'open' the black box, studies have thus tended to focus on controversy, in order to track black boxing as a process which closes down alternatives. Opening the black box has become a key aim, therefore, of studies of technology and culture.

Returning to Kline's definition, and moving outwards from a focus on the artefact, technology also refers, in his view, to the sociotechnical systems that manufacture hardware, to the information, processes and skills used to complete a given task, and to the broader sociotechnical systems of use, which bring together 'combinations of hardware and people (and usually other elements) to accomplish tasks that humans cannot perform unaided by such systems' (Kline 2003: 211). It is this latter, more encompassing notion of 'technology' that informs my argument; so, while I may focus on pieces of hardware, such as a fridge or a mobile phone, I am equally interested in systems of use that bring people and technologies together in specific ways. I would tack on to the end of Kline's tidy definition one more dimension, that is the Foucauldian sense of **'technologies of the self'** (Foucault 2000) – the practices of living within and through power structures that offer both freedoms and regulations (for a useful discussion, see Gauntlett 2002). This can be understood in terms of the codification of behaviour into *techniques* – repertoires of movements, actions, performances.

Andrew Murphie and John Potts (2003: 4) provide a similarly expansive yet nuanced set of definitions, noting that the term 'technology' was, in the nineteenth century, used to mean 'the application of a body of knowledge, or science, in specific areas' – notably the application of science to 'industrial' problems. So science and technology become split around the question of application, but also intricately bound together, as we'll see – though Mackenzie and Wajcman (1999: 7) note, 'it is mistaken to see the connection between them as one in which technology is one-sidedly dependent on science. Technology has arguably contributed as much to science as vice versa'. Murphie and Potts also note that second split, between technology and technique, which they take to mean the use of skill, both physical and mental.[2] Mackenzie and Wacjman concur, seeing technology having three layers of meaning: the artefact, surrounding human activity, and human knowledge. These three layers are woven together, so thinking about technology means thinking about the three simultaneously, and the myriad connections between them. And it is to thinking about the 'social effects' of technology that I now want to turn.

Technology in/and/as society

As Mackenzie and Wajcman (1999) set out from the start, the most common and also most problematic way of thinking about technology is what's known as technological determinism. This is a straightforward cause-and-effect equation: *technology produces effects in society*. These effects may be positive (in terms of technological progress) or negative (technology enslaves us, or deskills us, for example). Here technology is seen as separate from society, and as shaping it. People adapt to technologies that materialize, or 'land', from some asocial or precultural realm. Now, as we saw in the last chapter, the idea that things like scientific or technological practice take place outside the social is utterly untenable; as Mackenzie and Wajcman stress in the quote that opens this chapter, technology means *human-made* things (and connections between humans and things of different kinds). However, this does not mean that technology has no social effects – there can be no denying that, say, the mobile phone has changed the ways some parts of society operate. It's just to say that we shouldn't reduce our understanding of technology or society to the latter, passively experiencing effects propagated by the former. At the same time, taking a critical view on technological determinism must be mindful of the fact that it is a popularly available and widely held viewpoint; as such, we must also interrogate the reasons for its popularity (see Bell 2001).

Academics from a range of disciplines have been trying to find different ways of thinking about technology and society, in a pattern that in some ways parallels (and at times overlaps with) the moves described in the previous chapter to think about science and society (ideas like black boxing, already discussed, are used to explore science as well as technology). Setting aside technological determinism – though without denying its purchase as a discourse about technology and society – means doing a number of things differently; or, to steal Apple's strapline for the iMac, it asks us to 'think different'. As with science studies, part of the task here is to show that technology is absolutely social; that its invention, production, distribution, consumption are all social processes. Harvey Molotch (2003) makes this abundantly clear in his book *Where Stuff Comes From*, which focuses on the design of everyday artefacts, showing how the final shape of any object or product is the result of countless activities that all bear the imprint of society. Does such an insight mean that we should instead talk of social determinism – seeing society producing effects on technology? Of course not, for any simplistic one-way determinism is equally flawed. Nevertheless, researchers pondering this issue talk in terms of the social shaping of technology – Mackenzie and Wajcman's (1999) title – or of the social construction of technology (Bijker and Law 1992), in order to try to tip the balance away from simplistic technological determinism.

In fact, we need to talk of an assortment of social *shapings* of technology. Molotch (2003) reveals both purposive and happenstance moments in the design of the artefacts in his study; we also need – to tweak Latour's (1987) formula for lab studies – to *follow technologies around*, to track their development and their everyday lives, to see all the manoeuvring and negotiation between people and things as they work out a way to live together. And, as my fridge story shows, that living together also includes both elements preset into the technology – such as the functions of a fridge to keep food cold or frozen – and those 'double life' adaptations that turn fridges into message boards or child-minders. 'Where stuff comes from' needs matching with where stuff goes to and what becomes of it – and this is exactly the kind of tack that some researchers have taken.

Following technologies around

In the social construction of technology (SCOT) approach, this tracking is achieved in part by working to identify so-called relevant social groups or interest groups who have had a role in the way a piece of technology has turned out. Let's visit a well-known example from the literature: the bicycle. Ask someone today to describe or draw a bicycle and they will tend towards the same thing (though I always find them very hard to draw!) – a recognizable arrangement of frame, wheels, saddle, handlebars, and so on. In his exemplary SCOT-ish study of the history of the bicycle, Wiebe Bijker (1995) maps out the relevant social groups who influenced the development of the modern bicycle and shaped its form, also pointing to those machines deemed unsuccessful, and that have largely disappeared from modern bike shops (appearing instead, like kerosene lamps, in antiques shops). The bone shaker and the penny farthing have long gone, as has the tricycle (except for infants), and most bikes today are modeled on the low wheel 'safety bicycle' – in the language of SCOT, this has become the *stabilized* form of the bicycle. The bicycle has become black boxed, and we are blinkered from seeing it otherwise (for critique, extension and updating of this story, see Rosen 2002). Echoing the 'symmetry' of the Strong Programme in science studies discussed in the previous chapter, SCOT is as much interested in failures as successes; like science studies again, it works through detailed case studies, tracking the intricacies of social and techno-logical change. In Bijker's study, then, we see clearly that these bikes weren't merely foisted on a passive public who rode them without question; different groups, such as cycling clubs, women cyclists, as well as bicycle manufacturers, are all part of the story.

Also part of the story, of course, is the bicycle, and its many component parts; Bijker shows how the development of the tyre, for example, helped shift bicycling culture not only by reducing vibration – which suited some users – but also by increasing the bike's speed, satisfying sporting cyclists. And so we come to an important moment in thinking about technology and society: the moment when *things* were given a fuller place in the stories researchers told.

To explore this moment – without claiming in any way to be telling its 'origin story' here – I want to turn our attention to two essays by Bruno Latour (for an overview of his work, see Bell 1999a). Published in 1991 and 1992, both have richly evocative titles: 'Technology is society made durable', and 'Where are the missing masses?' The latter posits that sociology should pay attention to non-human artefacts – these are the missing masses that do all kinds of work in society, but which haven't been treated as part of society, and thereby have been ignored by sociologists. Latour's focus is on 'mundane' things, things like door-closing mechanisms, but his point is a big one: humans have delegated all kinds of tasks, tasks which keep society ticking over, to non-humans. So a door-closing mechanism does a job that people could, or maybe should: closing the door behind them. That we have developed an ingenious device to carry out this task suggests that people don't close doors, even if they're asked to (hence also automatic doors, revolving doors, and other similar solutions to the problem of non-door-closing humans). And in 'Technology is society made durable', Latour makes a similar point from a different angle, asking what distinguishes human societies from those of our near Darwinian relatives. The answer is **delegation**, again: human society is kept going by things, as people can't be continually doing the interpersonal 'face work' that keeps simian societies together. We produce things that enable our societies to live beyond face work, therefore – things like books or tools, which have a durability beyond direct interaction.

Ideas like these, and names like Latour's, are associated (rightly or wrongly) with another way of thinking about technology and society, known as actor-network theory, which I want to talk about now. Perhaps I should begin by saying that actor-network theory (ANT) is as much about science and society (and nature and society) as it is technology and society. As Mike Michael (1996) writes, it exists in an uneasy lineage with other ways of thinking, such as the **sociology of scientific knowledge** (SSK), even as it diverts away from SSK in its more expansive perspective on different people and things. ANT wants to take ideas like symmetry further still, by flattening any distinction between human and non-human 'actors' – so, if you like, bicycle tyres are to be considered as seriously as cycling clubs in understanding the development of the modern bicycle – and by trying to find ways of talking about things that are equally flat, non-hierarchical. To the outsider, this can lead to perplexity, as it is unfamiliar to use language in this way, and to talk with equivalence about people and things that seem, in many ways, extremely different. In explaining and exemplifying one of ANT's key ideas, *intéressement*, for example, Michael notes that it

> encompasses a variety of strategies and mechanisms by which one entity – whether that be an individual like Pasteur . . . a small group like the three biological researchers of St Brieuc Bay . . . or an institution like the Électricité de France . . . attempts to 'corner' and enroll other entities such as scientists, publics, institutions, scallops and electrons.
>
> (Michael 1996: 63)

This bringing together of apparently very different 'entities' – Pasteur, scallops, publics – has been labelled by John Law (1987) '**heterogeneous engineering**'. Such an approach allows us to keep in mind that something akin to techno-logical determinism does occur, in terms of what Hugh Mackay (1997) refers to as the constraining capacity of the physical or material. So, in an example eluded to by Michael, Callon's (1986) study of attempts to develop an electric vehicle in France, one 'entity' that refused to be 'enrolled' into the 'network' was the fuel cell, which didn't work as its designers intended. The cells were, as Michael (1996: 59) puts it, 'recalcitrant'.

The detail of ANT is beyond the scope of my discussion here (but see the glossary for an oversimplified description). And anyway, it's been 'broken' now, as Law (1999: 12) says. The important point to take from it is that our thinking about technology and society stands to learn a lot from including the non-human 'missing masses', from understanding delegation, and from seeing heterogeneous engineering at work in the multiple sites where technology meets people. Let's turn now to one way of thinking about that, in exploring tech-nology and something called 'everyday life'.

Technology and everyday life

It is probably worth beginning here by noting the flurry of interest, particularly from cultural studies, in everyday life (for an overview and some readings, see, Highmore 2002a, 2002b). In part this is to do with the 'rediscovery' of past theorists of everyday life, such as Michel de Certeau or Henri Lefebvre, and in part it reflects a turn towards the ordinary stuff of life and away from things that seem more 'spectacular'. As usual with cultural studies, of course, there's a lot of ambivalence about what 'everyday' means, too much to fully outline here. In fact, I concur with Maria Bakardjieva (2005: 38) when she writes that 'it would be an excruciating task to follow all the lines of reasoning drawn through and around everyday life in an attempt to resolve the debates still raging'. So, to cut to the chase, here's Ben Highmore's take on the meanings of the everyday:

> On the one hand it points (without judging) to those most repeated actions, those most traveled journeys, those most inhabited spaces that make up, literally, the day to day . . . [But the everyday is also] a value and quality: everydayness. Here the most traveled journey can become the dead weight of boredom, the most inhabited space a prison, the most repeated action an oppressive routine. Here the everydayness of everyday life might be experienced as sanctuary, or it may bewilder or give pleasure, it may delight or depress. Or its special quality might be its lack of quality. It might be, precisely, the unnoticed, the inconspicuous, the unobtrusive.
>
> (Highmore 2002a: 1)

One thing that's striking about this ambivalence is the extent to which it only sometimes touches discussions of technology and everyday life. What primarily interests some of the scholars whose work I will sketch in this section is the technology – and, it might be added, a lot of that work shares what Mackenzie and Wajcman (1999) call a 'soft determinism' in terms of looking for how technologies shape or reshape everyday life. Everyday life remains a given, taken for granted.

Let me give you a couple of examples, before turning to research that has, for my money, a more productive take on the relations between everyday life and technology. Claus Tully (2003) carried out research on young people's adoption of certain types of 'new' technology, and found that 'as they make use of technical artifacts, the everyday lives of young people change, as does their perception of society, because it is through the artifacts that relationships with others are organized' (444). While he wants to suggest that young people do have some agency in how they use these artefacts, noting also that 'they use them to produce new meanings are lifestyles' (455), this is a fairly uni-directional equation. Palen *et al.* (2001) looked at first-time mobile phone users, noting how they quickly accommodated mobiles into their day-to-day routines, even if they had predicted beforehand that they didn't need or want a phone. As they describe it, the mobile phone users in their study expanded the range of uses of their own phones as they became progressively embedded in 'mobile culture' – even changing their opinions about phone etiquette such as use in public and social settings. And even while they contest any simple 'impact' model in their study of domestic Internet usage, Ben Anderson and Karina Tracey (2001) end up concluding that Internet use doesn't take much time away from other, pre-existing domestic activities (except, perhaps oddly, cooking). As they put it, their participants 'are doing old things in new ways and finding that some of those new ways suit their lifestyles better' (473).

Offering much less deterministic perspectives come recent studies by Lehtonen (2003) and Bakardjieva and Smith (2001). The former tracks the adoption process, using the metaphor of trials – the trial of deciding a particular technological object is needed or wanted, the trial of settling-in with a new object that necessitates a reshaping of relationships with people and with other technologies, and the trial of disposal, of getting rid of things. His discussion of older forms of music-playing technology, such as audio cassettes, is especially resonant and evocative (see Chapters 4 and 7). The idea of a trial here, in Lehtonen's view, has a double meaning – of trying things out, and of finding them trying. Borrowing ideas from work on the 'biographies of things', Lehtonen maps this life-path as an initial infatuation followed by progressively 'normal' cohabitation, and ending up in taken-for-grantedness, abandonment, redundancy and disposal. Arguing that adoption or domestication is too often 'black boxed' or taken as self-evident, Lehtonen instead describes the process as a series of trials 'for two main reasons: first, because of the *openness* of this conceptualization; and, second, because of its tendency to direct attention to tensions and *dynamic relationships*' (381). Importantly he notes that trials produce knowledge, ways of using technologies and of thinking about them. As he concludes, this analysis 'makes it possible to locate an unpredictable range of heterogeneous forces meeting each other – a *multiplicity* – where surprises can occur and humans and nonhumans mutually change their qualities and capabilities' (383).

Bakardjieva and Smith (2001) make a similar point in their study of 'ordinary', domestic Internet use (see also Bakardjieva 2005). New social practices generated by use offer the possibility to (socially) shape the Internet, for example by generating new forms of content: 'Ordinary users have performed important signification work contributing to the public definition' of the technology, identifying 'empowering potentialities' (80) – even though there may be a question mark hanging over the translation of these potentialities into anything more concrete (though they suggest that things like blogging might effect this translation). Even if that is irresolvable, they borrow from Lefebvre, one of the kings of everyday-life writing, to note the value of 'the critique of the real by the possible' (81). Ultimately, what we must take from studies like these is a point made some years previously by Constance Penley and Andrew Ross (1991: xiii), in their discussion of 'technoculture' as 'located as much in the work of everyday fantasies and actions as at the level of corporate or military decision making'. In other words, everyday life is a vital site of enquiry for anyone with an interest in understanding 'the dangers *and* possibilities' (xii) stirred up when those heterogeneous forces that Lehtonen spoke of, meet. And where they meet is often precisely in the spaces of everyday life.

Among the most significant but also ambivalent of the spaces of everyday life is the home, although Bakardjieva (2005) rightly notes that the two things should not be straightforwardly conflated or collapsed together. A whole host of researchers has spent time exploring what goes on when technology comes home. Connecting home and technology brings up the idea of domestic technologies – back to fridges again – and all the other technological artefacts we use at home. There are, it has been argued, two forms of domestic technology: one is concerned with domestic *work*, like vacuum cleaners or tumble driers, and the other with domestic *leisure*, such as televisions and games consoles. Admittedly, some technological objects can move between these categories – the home computer being an obvious candidate, used to do homework or for working from home, but also for playing games or chatting to your mates on email – though the importance given to these activities is often waited judgementally in favour of 'productive' uses (Mackay 1997). But it's impossible to talk about domestic technologies of whichever type without thinking about household or family politics, things like gender and generation. As Rita Felski (2000) has argued, we must pay special attention to the gendering of everyday life, and remember that 'home' is more than a spatial location; it is a container of metaphors, many of these also powerfully gendered.

Technologies of gender

Work on gender has been incredibly important for considering the mutual shapings of technology and society, in exploring the technologies of gender and the genders of technology. Domestic technologies have been a key focus of this work, and to get a feel for arguments about gender, home and technology, I want to offer brief sketches of three more-or-less representative examples, all sampled in the reader *The Politics of Domestic Consumption* (Jackson and Moores 1995). Cynthia Cockburn's (1995) 'Black & Decker versus Moulinex' explores how the gendered division of domestic labour is reproduced through domestic technologies – men use tools to do DIY, women use utensils or appliances to do housework. Moreover, she picks out through empirical research the gendered assumptions about skill and knowledge (techniques) that keep women away from certain technological tasks. These techniques are emblematized in 'the shed at the bottom of the garden' – the shed of male tinkering, fixing and making things (on men's sheds, see Thorburn 2002; on sheds and boffins, see Chapter 5). Cockburn's respondents talk about the shed (like the garage) as an exclusionary space, but also a space for the transmission of technique from father to son. The shed is, thus, a kind of black box for women: they are expected to know nothing about its inner workings, only about its inputs (men, cups of tea, etc.) and outputs (a repaired blender, a hand-made spice rack, etc.).

Ann Gray (1995) looks at the video recorder (VCR) in the home, also with a focus on technique and on black boxing – the gendering of the skill and knowledge of operating the VCR (especially its more complex functions, like programming). However, Gray detects a distinction between a kind of intentional ignorance on the part of some women, who recognize that knowing how to do things means then *having to do them* – what she calls a 'latent servicing function' (235) – and other women, who are excluded from the techniques of the VCR. As Gray writes, this exclusion has negative repercussions for these women:

> They feel stupid because of their lack of knowledge in this area. That [this] can be accounted for in terms of material restrictions – having particular domestic duties to perform rather than being able to sit down and study an instruction manual and its application – is then turned back on the women, often in their own consciousness, as a presumed basic inability to understand technical things.
>
> (Gray 1995: 235)

Even if this may be, as Gray suggests, a 'transitional stage' that is passed through once the technology loses its scary (but special) newness and technologicalness, and becomes mundanized, the broader impact of reinforcing the perception that women are 'no good with technical things' maintains, to use Gray's colour-coding of technology, a blue box as well as a black one.

Of course, the technologization of domestic work should act as a powerful counter to this notion of women's lack of technological abilities; programming a washing machine is no less taxing that programming a VCR, after all. Judy Wajcman (1995) turns her attention to the broader relationship between technology and domestic work, or to the technologization of domestic work. This is an issue where a kind of technological determinism might be desirable – if new technologies could produce effects in terms of transforming housework, notably through the idea of being (domestic) 'labour saving'. But as Wajcman notes, this hasn't happened. Domestic technology has been, like the fuel cells Michel Callon studied, recalcitrant when it comes to reducing the time and effort spent (largely by women) on housework. While this isn't the result of a technical failure of the technology, but caused by things like rising standards of cleanliness (see Shove 2003), it nevertheless serves as a useful reminder of the tension between social and technological shaping.

Wajcman also considers futurological imaginings of high-tech home life, noting its perpetuation (even heightening) of gender inequalities; the 'smart house' of the future, she remarks, will require extensive *programming* – programming which, to echo Gray's VCR study, 'may enhance men's domestic power' (223). If the smart house's technological solutions to issues of domestic work and gender politics are so woefully lacking, then what about social solutions? Wajcman surveys some past ideas, such as the socialization or collectivization of housework, but notes the external factors inhibiting their development (these include the ideology of home ownership, the use of material possessions to signify wealth, and so on). Ultimately, she concludes that gendered meanings are deeply encoded into domestic technologies – through the design and manufacturing process, for example. The task is to decode those meanings, to expose the relations and inequalities they perpetuate.

This discussion of the smart house chimes with other studies of domestic futurology and its genderings. Ann-Jorunn Berg (1994) notes the way that smart houses prioritize high-tech gadgetry at the expense of décor and style – other than the 'masculine' style of minimalism (see Holliday 2005a). As Berg puts it, 'the smart house is no home' (177). Katja Oksanen-Sarela and Mika Pantzar (2001: 212), meanwhile, ask other questions of the home life imagined in smart houses: 'what about people for whom home is no paradise? They are not present in the stories of the future . . . [Nor are] incompetent users or "lazy people" who are not willing to invest their time' on becoming as smart as their house, or at least smart enough to dwell comfortably within it. Part of the reason why smart houses seem such hard work, they continue, comes from what they call 'utilitarian individualism' (214) – a desire for smartness, for self-improvement, based on the idea that the 'proper' use of technology should be productive or pedagogic. Such a logic has long haunted playful uses of technologies, as the tension between the home computer as a schooling device and as a games machine showed (Mackay 1997).

Looking at recent visions of smart living in the USA, including that provided by Microsoft boss Bill Gates in his book *The Road Ahead* (1995), Lynn Spigel (2001) detects a 'curious blend of baby boom nostalgia and "gee whizz" futurism', which she reads as 'both symptoms of a more prevailing anxiety and confusion about the contemporary world' (407). She notes how Gates depicts technology as a servant and companion, conjuring virtual entertainments and handling domestic chores in a smart way, which here means unobtrusively, as well-trained human servants did in the past. This leads Spigel to think about the issue of class as it is encoded in images of smart houses; images which, she writes, show us 'how to use new mobile technologies like fax machines, mobile phones, and the internet to *conserve* – rather than transform – middle-class values of family life and home' (403; my emphasis): no chance of Wajcman's socialist communal laundry there, then. Gender relations are to be transformed by smartness, but in a retrograde direction, since 'new technologies can reverse the "damage" done to families by women's liberation, and particularly by women's entry into the workforce' (ibid.). The obvious question becomes, therefore, *smartness for whom?*

The relationship between social class and technology has received less attention than that of gender. Spigel's observation about smart houses and middle-classness brings this issue neatly to the fore, raising the broader question: do technologies have class? Given that we have already discussed the ways they have gender, and referred to the nationality of fridges, we can assume an affirmative answer, and begin to look for examples of the classing of technology (and its corollary, the technologizing of class). Charlotte Brunsdon's (1997) work on television satellite dishes in the UK provides a suitable starting-point. When the only widely available way of receiving non-terrestrial TV channels in the UK was via a satellite dish installed to the exterior of the house, a clearly classed discourse of snobbery and taste was produced. Having a satellite dish marked your household as couch potatoes, as people who watched too much telly – and in Britain this has a powerful class connotation linked to an old idea about productive leisure (Rojek 2001). Now that non-terrestrial programming can be received at home through less visible devices such as set-top boxes or cable, and as the kinds of programming available have changed, so this class equation slowly shifts, though it is still a powerful residuum.

Research on mobile phones also has something interesting to say about class, and about the shifting class connotations of the phone as it moved, in the UK at least, from 'yuppie' accessory to schoolyard must-have (its yuppie significations being passed on, in some ways, to the palm pilot and then the Blackberry; see Bell 2005). There's a similar kind of snobbery at work that views mobile phone use as a kind of dependency, and devalues the kinds of talk on mobiles as gossip, even gibberish (Palen *et al.* 2001). The rapid adoption of text (SMS) messaging among mobile users – or at least some kinds (and some classes) of mobile users – has rejigged the class signification again, in complex ways, as shown by 'moral panics' in the UK over the effects of 'textspeak' (or txtspk) on standards of literacy (Bell 2001).

In terms of the technologization of class, there has been some important work on class-based barriers to new technologies, whether these are economic, educational or social (or a combination of all these, and more). The production of a 'digital underclass' as a result of differential access to computers and the Internet is probably the most prominent line of enquiry, and also the most prominent area of policy intervention, as governments work to flatten these hierarchies (Loader 1998). As these sketches hopefully suggest, then, there are important classings at work here; as with work on gender and technology, these are revealed to be both encoded in the development, form and selling of technologies – class barriers can be erected by design, by advertising, by price, and so on – and also in the decoding of artefacts as they continue to develop a 'social life'.

The ambivalent twinning of futurism and nostalgia, of high tech and pastoral, that Spigel spots in Bill Gates's house, leads us towards another question: given the black boxed 'inevitability' of the increasing 'smartening' of home life, even if not to the super-high tech extent imagined by futurologists, what other ways of living might be imagined or experienced that resist 'smartness'. There's a commonly circulating set of anxieties about what Andrew Ross (1994) calls 'the new smartness' – the anxiety that we humans will be outsmarted by smart devices, that technological determinism will take the form of a battle between human and machine. This is a battle which, judging by dystopian sci-fi, we seem to figure we will lose (see Kuhn 1990). One response to this, which has a long history, is resistance through rejection, often talked about as **Luddism**, in reference to industrial unrest in nineteenth-century Britain over the increased technologization of work. Luddism's anti-technology stance has been reworked in recent times, most notably in the USA's neo-Luddite movement (see Robins and Webster 1999). Associated with people like Kirkpatrick Sale, who likes to stage the smashing of computers, or (wrongly but nevertheless routinely) with Theodore Kaczynski, the so-called Unabomber, with his infamous manifesto and decidedly *anti-smart* shack in the Montana wilderness,[3] neo-Luddism conjures pre-technological life as the best post-technological prospect: 'a return to nature and what are imagined as more natural communities' (Robins and Webster 1999: 61). Of course, the question of who does the housework in a Neo-Luddite world finds the answer the Wajcman won't like to hear, given the movement's ideas about family life and family values.

Even though it is a fringe movement, some of neo-Luddism's ideas, critiques and solutions have a broader resonance in contemporary culture, expressed for example in assorted moves towards more 'simple' ways of life – where simple and smart have their values inverted, the former positively invested and the latter seen negatively. Lots of people want to *slow*, if not halt, the rapid churning of new technologies; they want to question the (deterministic) impacts these have on society; they want less smartness, because smartness in things is starting to scare them (until they can domesticate it, of course). Less speed, less change, less newness, less technologicalness. On a recent radio panel show, this sentiment was made clear, with nice irony, in a skit in which the contestants were asked to brag to each other about their mobile phones, continually upping the ante. The final brag, with beautiful circularity, went something like this: *my mobile is so amazing, it comes with a house attached to it.*[4]

Thinking about technoscience

There's something uncomfortable about the way I've framed this chapter and the previous one. I've done something that I shouldn't have: I've split my discussion between science and technology. It may produce a nice pair of chapters, but things aren't that neat outside this book. As I did note earlier in this chapter, the long and commonly held distinction, of science as 'pure' and technology as 'applied', has vanished, and with it the relative valuing of scientists and technologists. In part this is due to the legitimation crisis in science, and the idea that science needs to be more useful (and at the same time less dangerous). It's also due to the blurring of any distinction between the concerns and practices of the professions of science and technology, and the realization that their interrelationships are manifold and multidirectional. As Mackenzie and Wajcman (1999) remind us, much scientific work is enabled by technology (such as computers), reversing the past-held idea of dependence which ran the other way.

Such a mixing and blurring has led some commentators to focus their ideas on a 'new' hybrid, which they call technoscience (for a critical view from a cultural studies angle, see Reinel 1999). Steven Best and Douglas Kellner (2001) describe the co-evolution of science and technology, and their morphing into the conjoined technoscience, as a series of implosions, not limited to the intersections and cominglings of science and technology, but beyond these to also pull in capitalism, the military, culture, everyday life:

> Technoscience has helped to generate a world of glass, steel, plastic, highways, synthetic fibres, and chemicals, gadgets, and new forms of culture such as cyberspace and virtual reality (VR), as it has also genetically engineered life forms and reconstructed 'the natural' itself. Technoscience . . . manufactures a surfeit of industrial and household products, the infrastructure for the media/computer/biotech transformations of our era, and contributes greatly to the construction of rapidly changing social and natural worlds.
>
> (Best and Kellner 2001: 101)

But technoscience is not just a quicker way of saying technology-and-science, nor only about the collapsing together of two previously bifurcated intellectual traditions, practices and professions. As Aylish Wood (2002) writes, in her book on films of technoscience, the term has come to stand for the 'whole variety of factors that influence the outcome of attempts to create knowledge or objects' (3). She quotes Donna Haraway's (1997) nicely heterogeneous list of what kinds of thing count as a 'being' in technoscience: 'a textbook, molecule, equation, mouse, pipette, bomb, fungus, technician, agitator, or scientist' (Haraway 1997: 68, quoted in Wood 2002: 3).

Haraway is seen as an important figure in ideas about technoscience. Introducing the term and its attractions, she writes that

> technoscience is a form of life, a practice, a culture, a generative mix . . . The world-building alliances of humans and nonhumans in technoscience shape subjects and objects, subjectivity and objectivity, action and passion, inside and outside in ways that enfeeble other modes of speaking about science and technology. In short, technoscience is about worldly, material-ized, signifying and significant power.
>
> (Haraway 1997: 50–1)

By emphasizing power, Haraway asks that we 'cast our lot', that we find ways of being in a technoscientific, power-infused culture. We cannot, she says, 'remain neutral', but must 'squirm, organize, revel, decry, preach, teach, deny, equivocate, analyze, resist, collaborate, contribute, denounce, expand, placate, withhold' (51) – again, a heterogeneous broth of different responses. Now, commentators such as Birgit Reinel (1999) or Patricia Ticineto Clough (2001) argue that the emerging, equally heterogeneous 'field' of cultural studies of technoscience has the capacity to help us think how we might 'cast our lot', and not just in terms of academic debate: 'cultural studies of technoscience offer the possibility not only to broaden the scope and range of cultural studies' fields of enquiry, but also to actively shaped the politics and directions of [contemporary] western technoscience' (Reinel 1999: 181). Such a project means rejecting determinisms, no matter which direction they hail from – though without forgetting the enormous popular appeal and circulation of deterministic thoughts and images. It means being attuned to all that heterogeneity, all that complexity. In fact, it means *inviting* complexity, as the only way of casting our lot in this complex technoscientific world.

Further reading

Bakardjieva, M. (2005) *Internet Society: The Internet in Everyday Life*. London: Sage.

Bell, D. (2001) *An Introduction to Cybercultures*. London: Routledge.

Best, S. and Kellner, D. (2001) *The Postmodern Adventure: Science, Technology, and Cultural Studies at the Third Millennium*. London: Routledge.

Bijker, W. and Law, J. (eds) (1992) *Shaping Technology/Building Society*. Cambridge MA: MIT Press.

Jackson, S. and Moores, S. (eds) (1995) *The Politics of Domestic Consumption: Critical Readings*. London: Prentice Hall.

Lehtonen, T-K. (2003) The domestication of new technologies as a set of trials, *Journal of Consumer Culture*, 3(3): 363–85.

Mackenzie, D. and Wajcman, J. (eds) (1999) *The Social Shaping of Technology* 2nd edition. Buckingham: Open University Press.

Michael, M. (1996) *Constructing Identities*. London: Sage.

Murphie, A. and Potts, J. (2003) *Culture and Technology*. Basingstoke: Palgrave Macmillan.

Shove, E. and Southerton, D. (2000) Defrosting the freezer: from novelty to convenience, *Journal of Material Culture*, 5(3): 301–19.

Sismondo, S. (2004) *An Introduction to Science and Technology Studies*. Oxford: Blackwell.

Notes

1 I remember some television programme saying that in very cold places, people use fridges to keep things from freezing, to keep them 'warm'.

2 It was at this point in writing the first draft that the fridge arrived! But it was a short-lived arrival – when it was unpacked, it was found to be damaged, and so it had to be taken back. Maybe by the time the book is published, there will in fact be a new fridge here to party around.

3 See the series of short personal accounts in *Science, Technology & Human Values* (Benson 2001; Restivo 2001; Strum 2001) on the Unabomber investigation, into which members of the US Council for the Society for Social Studies of Science (4S) were drawn, as experts but also as suspects. Restivo and Strum both note that Kaczynski was misappropriated as a posterboy for technocriticism or Neo-Luddism, mainly thanks to the tenor of his manifesto, published in *The New York Times* and still widely read, thanks to the Internet, by followers of a certain 'extreme' strain of 'technomillenarian' thinking (see also Chapter 6).

4 It was on BBC Radio 4, and although I can't remember which show or when, the person who made that final brag was, if my memory serves me correctly, Jeremy Hardy.

SCREENING (AND SINGING) SCIENCE AND TECHNOLOGY

The relation between science and fantasy is a very complicated one – facts and fictions cannot be segregated neatly into different compartments, but weave into one another in very strange ways.

(W.J.T. Mitchell)

My aim in this chapter is to explore some sites in and through which science and technology are represented in popular culture. In particular, I am interested in looking at places where audiences constructed as 'non-specialist' (which means non-scientists in the main) access information, images and ideas that help them think about science and technology. What resources are made available to 'lay' audiences – to 'ordinary people' – to make sense of science and technology? What kinds of representations are made popularly accessible, in what formats and on what platforms, and what kinds of messages do these transmit?

Answering these questions, and countless others that they catalyse, necessitates a narrowing of focus and reach. There are so many possible routes to take, so many examples that could be worked through, that this whole book and many others would soon get filled up. So, this chapter is necessarily partial, and focuses on two main stages for the representation of science and technology (though, of course, I will allude to others along the way): film (especially fictional, narrative films, mostly those labelled science fiction) and popular music. I have chosen these as each offers different responses to and different frames for representing and thinking about science and technology.

Science and technology at the movies

The first representational form I want to consider, then, is film: or, to be more precise, narrative mainstream fictional movies of the kind produced by Hollywood, mostly those conventionally labelled 'science fiction' – itself a contested term (Wood 2002). I want to trace two approaches to understanding (science fiction) cinema, which come from different intellectual traditions, and produce different readings of how these films are produced and consumed. One approach is rooted in the traditions of film studies or film theory, the other emerges from work on the public understanding of science. By and large, we can make a theoretical and methodological distinction between these two strands – though part of what I want to do here is to show how they have been or might be brought together – which, to simplify in the extreme, goes like this: analyses from the film studies tradition start with the film itself, and ask how it represents issues to do with science and technology; work using the public understanding of science (PUS) approach starts with the science or technology, and then asks how these are represented in films. This distinction might seem so subtle as to be almost meaningless, but I hope to show in the discussion which follows how this difference leads to quite radically different ways of thinking about film-making and film-watching, and about the practices and meanings of science and technology.

Both approaches, however, share a big question: is science and technology – however it is embodied in the particular movie under analysis – shown in a positive or negative light on screen? While scientists have argued recently that movies are responsible for a widespread popular hostility towards science and technology, thanks to a 'dystopian turn' in sci-fi (Gregory and Miller 1998), this is an oversimple reading of the effects of film-viewing. As we shall see shortly, this oversimplification is symptomatic of PUS approaches, which often fail to consider the 'cultural work' that film audiences participate in, and the **intertextuality** that locates filmic images in relation to other resources (news media, science education) drawn on by people in order to work through what science and technology 'mean'. However, criticizing the Manichean analysis of movies as straightforwardly offering *either* positive *or* negative representations of science and technology should not mean dismissing the popular circulation of this central idea: that films do produce effects in terms of changing (or reinforcing) how their audiences think. Neither should it mean ignoring the centrality of stories about 'good' versus 'bad' science and technology as they circulate in popular culture, including popular film. As we shall see later, more recent analyses of cinematic representations of science and technology consider a greater variety of 'reading positions' and responses to these stories (as well as exploring how the stories themselves carry more complicated or ambivalent messages about science and technology; see Wood 2002). Popular critical reviews of films – those carried in newspapers, for example – routinely offer precisely that kind of 'common-sense' reading of films; so, while film theory is very useful in terms of complicating our readings, we should not lose sight of the ways films are talked about and thought about not only by non-specialists in terms of science and technology, but also non-specialists in terms of film theory.[1] So, for all its shortcomings, it seems appropriate to start with the PUS approach, which black boxes any of the complexities of film theory, and asks quite simply: what are films saying about science and technology?

Public understanding of science fiction

Analyses of the public understanding of science (PUS) are numerous and diverse, but they seem to share at their heart a pro-science desire: to explore how science and technology are currently represented to non-scientists, across a wide range of domains, and then from that exploration to devise a programme to better represent science and technology – to enhance public understanding of science and technology in equally pro-science ways. So, while Jane Gregory and Steve Miller (1998: 52) are right to say that 'popularizing science may not necessarily make science more popular', in that greater public access to and scrutiny of science and technology may do more to fuel antiscience sentiments (see Chapter 6), there is an underlying rationale to PUS-based approaches to popular film that starts with the idea that audiences consume images of science and technology quite straightforwardly, or transparently, and that the task is therefore to make those representations 'better' – more accurate, more balanced, more pro-science.

This is part of the much broader project of PUS, which seeks to understand and intervene in the many ways that non-scientists experience science and technology (or ideas and images about science and technology), founded on the assumption that a better public understanding of science and technology is good for people, good for society – and good for science, which is currently 'misunderstood' or not understood (for a critical overview, see Allan 2002; Irwin and Michael 2003). The second, equally important assumption underlying the PUS agenda is precisely that: that science *is* misunderstood or not understood, and moreover that it is represented negatively in popular culture. As Stuart Allan (2002: 45) puts it, 'scientists, it would seem, have an image problem'.

In order to understand this image problem, we need to backtrack a little, and make some very general comments about the place of science and technology in contemporary culture, already laid out more fully in Chapter 2. This issue centres on the development of a separate cultural domain called 'science', a domain of particular practices carried out by particular people – scientists (the parallel argument can be made for technology and 'technologists'). This story is all about knowledge, about who 'owns' knowledge, whose knowledge is valued – and how those kinds of judgement become institutionalized in a society through things like education, careers, politics. In the process, science has black boxed itself, in an attempt to protect its status and its truth-claims; so, those people invented or constructed as 'lay people', as non-scientists, are kept out of the laboratory. But, of course, as we saw in Chapter 3, black boxing produces contradictory effects: if non-scientists don't have access to science, they can't be expected to understand it, and so how can they trust it? As the so-called legitimation crisis in science has taught us, if people can't understand and don't trust science, this puts science in a tight spot, so long as it relies on popular support, for example, in order to maintain political support and concomitant economic support (see Chapter 2). Hence the drive to enhance the public understanding of science, in which 'science' and 'public' are conceived as separate realms, both ignorant of the other. Now, while things are certainly more complicated than that – as Alan Irwin and Mike Michael (2003) show us, for example, even terms like 'science' and 'public' require some careful handling – this idea, that the public's understanding of science is in some way deficient, and that this is to the detriment of scientific practice and progress, casts a long shadow over the issue of representing science and technology.

As such, popular representations of science and technology have attracted academic scrutiny, including that coming from researchers with a broad PUS leaning. As I said at the start of this chapter, one way of typifying these analyses is that they start with the science and work to track how it has been represented in their chosen texts. Put another way, this means they tend to leave the science itself unscrutinized or uninterrogated. I want to consider here two examples of this kind of approach, which I take as more or less typical. The first is literary rather than filmic, but nevertheless produces a way of mapping representations of scientists resonant with popular film: Roslynn Haynes's (1994) *From Faust to Strangelove*. The second is a trio of papers by Robert Jones, published in the journal *Public Understanding of Science* (Jones 1997, 1998, 2001) – a journal which has also published Haynes's continuing work on literary images of scientists (2003) as well as other recent papers exploring aspects of filmic representation, including portrayals of female scientists, different generic conventions, and changing depictions of particular fictional scientists such as H.G. Wells's Dr Moreau (e.g. Flicker 2003; Jorg 2003; Weingart 2003).

Establishing the location of her work within the PUS tradition by writing that there is a need to 'confront the widespread, often unacknowledged, fear of science and scientists in Western society' (1994: 4), Haynes produces a broad-brush historical typology of Western literary depictions of scientists that stretches back to the Middle Ages. Across this vast time range she is able to detect a recurring group of six archetypes or stereotypes:

- the alchemist
- the stupid virtuoso
- the unfeeling scientist
- the heroic adventurer
- the helpless scientist
- the idealist,

We can sketch each of these in the following cartoons: the alchemist is obsessive in his (one key point from Haynes is that literature's scientists are only extremely rarely women) pursuit of obscure knowledge, often tinged with maniacal intent. The alchemist uses science to transform matter and so, as Haynes (1994: 3) notes, has more recently morphed into a kind of bio-alchemist, 'producing new (and hence allegedly unlawful) species through the quasi-magical process of genetic engineering'. The stupid virtuoso is brilliant but profoundly asocial – the geek or nerd who is so immersed in science he doesn't understand 'normal' everyday life. This 'asocialness' is, in fact, an often-used short-hand to perform 'scientist'. In its British lineage, this is the boffin (see Chapter 5), the unfashionable, ill-mannered or ill-at-ease misfit. As Haynes sketches them in her later essay, these boffins 'wear unmatched socks [and] never remember to cut their hair' (Haynes 2003: 248). This stereotype is often seen as far from the harmless, bumbling boffin, however, in that he is unable to foresee the social consequences of his scientific actions. He is in some ways similar to the unfeeling scientist, though this latter category is invested with a more romantic or heroic quality, having suppressed his feelings in the interest of scientific pursuits. Unable to find time to form human relationships, and devoid of emotion, the unfeeling scientist is a powerful and recurrent stereotype of the negative individual and social effects of science. Mathematicians, Haynes notes, are often portrayed in this way (recent cinematic examples include *A Beautiful Mind* and *Good Will Hunting*), but it was atomic scientists who first embodied the 'inhumanity' of this type. The heroic adventurer emerges in key moments when science is valued positively – the scientist is a saviour (or potential saviour), and often an inventor. While Haynes notes both the rarity of this type and its potentially sinister slippage into power-crazed superman mode, science is still represented as heroic in a great many fictional depictions: in later representations, science may be figured as simultaneously problem *and* solution, with conflict between different scientists (and sciences) staged and with the heroic scientist triumphing over other types (Wood 2002).

The helpless scientist, like his unfeeling and stupid lab-mates, fails to think through the implications of his scientific practice. He literally or figuratively births a monster which then runs amok. This idea, of science 'unleashing' monsters (or creating them), which it then can neither control nor destroy, is one of the most widely-circulating and prominent popular portrayals of the dangers of unchecked science. It has a mutant relative in the idea of science being taken away from the helpless scientist and used to demonic ends by other powers (governments, corporations), as we shall see. In this case, science is short-sighted but morally neutral (though of course moral neutrality is turned into something negative – amorality, like asociality, here restating science's problematic black boxing). Lastly, the idealist is a second positively charged stereotype, although his idealism is often shown to be thwarted by the context in which he is made to operate. In all six characters, then, we see a restatement of a split between science and society, where such a split is overwhelmingly seen to produce negative consequences: science without social, moral and human values is a dangerous thing.[2]

In her later reworking, Haynes (2003) redivides her stereotypes into seven categories, where the scientist is shown to be either an evil alchemist, or noble, foolish, inhuman, mad, helpless or an adventurer – but her basic point remains the same: scientists are overwhelmingly represented in popular narratives in negative ways. More importantly, Haynes writes that while they may seem overly simplified, 'we cannot root out these images without acknowledging their degree of veracity and [without] heeding their message'. So, while PUS-informed critics argue that fictional representations do more harm than good (e.g. Lambourne 1999), it is more important that we do as Haynes says, to acknowledge that these ideas and images do indeed have enormous popular resonance, that people come to them armed with a variety of prior ideas and images, some of which may be contested, others confirmed. The stories that we tell about science and technology emerge from different domains and different traditions, but they are no less important for being 'fictional' (see Wood 2002 on the blurring of such categories as 'fact' and 'fiction').

In a similar, if narrower and more detailed vein, Robert Jones has written about assorted British films from 1945 to 1970, exploring different facets of their representations of science and scientists, which he classifies into three principal stereotypes: scientists as artists, as destroyers, or as boffins. In the earliest of his three papers, Jones charts the rise of the archetypal boffin, who emerges in films about World War II (such as *The Dam Busters*, a biopic about inventor Barnes Wallis), and can be seen to be both heroic and a social outsider, obsessive about his work but innocent of the social context and social implications of science. In this way, Jones's boffin combines a number of Haynes's archetypes, and his analysis of *The Dam Busters* shows how the depiction of a single scientist's career and life can work through different variants of the boffin typology. Moreover, given the wartime context from which these movies draw their inspiration, a recurring motif is interaction – often difficult interaction – between science, the state (civil servants) and the military. This is an important point, in that subsequent cycles of films have equally explored interactions like these, with other agencies, showing their important impacts on the uses (or abuses) of science.

Jones concludes that the egg-head eccentric is nevertheless a sympathetic, even affectionate portrayal of scientists, while their connection here to victory in the War casts them as heroes. Elsewhere, however, he finds on the movie screen depictions of evil boffins, notably in the form of the James Bond villain. The shift from the war film to the spy film also keeps the heroic boffin in role, with Bond's lab-mate Q providing the archetype. Alongside the boffin, Jones tracks a set of representations that portray the scientist as a kind of 'artist'. Both groups are social misfits, both obsessive and both, Jones argues, problematically located in terms of the class system. While artists are shown as subjective and emotional whereas scientists are coldly rational, Jones suggests this places them in similar 'outsider' positions. Although Jones concludes that film-makers may use this scientist-as-artist model in order to explain scientific creativity, which might otherwise prove quite uncinematic, this confluence can also be read as a twist on the 'heroic scientist' type described by Haynes. Both artist and scientist are seen as uniquely gifted, complicated and eccentric geniuses – but also to be producing valuable work, even if it cannot readily be understood by 'ordinary people'. Making science like art, therefore, produces a further set of mystifications, keeping science (and art) bracketed off from 'ordinary' society.

In his later paper, Jones moves on to consider how cinematic narratives sometimes carry a critique of science, for example in terms of the arrogance of scientists, which blinds them to the social impact of their work. His analysis leads him to catalogue three main ways that the films he discusses *reflect* (an important word we will return to later) concerns about science in society. These concerns are about the use and control of science, scientists' lack of social responsibility, and – interestingly inverting the view of scientists as unemotional and inhuman – what Jones (2001: 374) calls 'scientists' lack of objectivity' (their tendency in films to be jealous, secretive, avaricious, and so on). As he says, the first and third concerns are seemingly at odds: scientists 'are expected to be uninfluenced by their emotions in the pursuance of their work, but this very lack of emotional involvement may mean that they are insufficiently concerned with how their research impacts on society' (375). And Jones ends his review with a final important yet contradictory point that we shall revisit: that science is often depicted as both problem *and* solution – there is bad science, but the antidote is good science.

While Jones's insightful triad of papers gives us some useful exemplification of filmic representation, his approach is typical of the pro-PUS camp, in terms of its inability to think theoretically about *film*: film is taken straightforwardly to reflect social issues and concerns. But, as Megan Stern (2003) cautions, we need to attend to the polymorphousness and the intertextuality of these representations. Films, like science, don't operate in a vacuum, and neither should our analyses. So, while Warren Wagar's *Science Fiction Studies* review of Haynes ends by saying her book is 'free of postmodernist philo-babble ... and mercifully oblivious ... to such technical lit-crit concerns as form, genre, and semiotics' (Wagar 1995: 5), I would rather call for us to *make use of* postmodernist philo-babble and technical lit-crit terms, if they are productive for our understanding of how popular culture 'texts' produce particular stories about science and technology, and how those stories are 'read' and made sense of not in isolation, but in that multitude of intertextual networks that comprises everyday life.

Film theory goes to the laboratory[3]

In her analysis of recent movies about technoscience, Aylish Wood begins with the idea that:

> Images of science and technology from the late twentieth century have long left behind the idea of the gentleman scientist working away in splendid isolation on his grand idea in his castle, or cellar, or even occasionally his elite community. What can be found, instead, are a series of images in which the outcome of a scientist's work is that which emerges from a complex network of influences, some constricting, and others unexpectedly productive.
>
> (Wood 2002: 1–2)

These films, she goes on to argue, 'refuse simple dualisms between good versus bad' (109) in their depictions of science and technology; neither are they straightforwardly technophobic or technophilic, and nor do they separate science from society. This more complex articulation of 'networks' of technoscience is rooted in a different approach to film than that associated above with the public understanding of science. It is based upon exactly the kinds of theoretical traditions whose absence in *From Faust to Strangelove* Wagar celebrates: Wood draws on an equally heterogeneous network of theory in order to think through a diverse programme of mainstream films, ranging from *Gremlins* to *Junior*, from *Nell* to *GATTACA*.

In moving from the PUS-centred approach towards analyses emerging more from a film studies or film theory direction, it is useful to turn to Annette Kuhn's (1990) discussion of different ways that science fiction films have been theorized. This overview provides us with a catalogue of potential analytical positions, and will hopefully help us understand the complexity of the cultural work of representing science and technology – something that some of the previous accounts fail adequately to accommodate. While Kuhn (1990: 10) is keen from the outset to stress 'the interdependence of, and the flow and exchange between, the varied theories and methods', she nonetheless identifies five main approaches, which I will list here before attempting to explain and exemplify each in turn.

- The first is analysis that sees movies as straightforward *reflections* of social attitudes and issues, whereby film is treated, as she puts it, like 'sociological evidence' (Ibid.) – this is the kind of analysis offered by many PUS-based researchers, as shown above.
- The second reads films *ideologically*, looking for the 'hidden meanings' that the movies contain and transmit, drawing on theories of ideology (often associated with Marxist cultural theory).
- The third strand draws on *psychoanalytic* theory, usually based around Freud or Lacan, to understand how films 'voice cultural repressions in "unconscious" textual processes . . . like dreams' (Ibid.) – again, movies have 'hidden meanings', and here these are about primal fears and feelings.
- Next are analyses that foreground something significantly neglected in PUS-based approaches as sketched above: an understanding of how *audiences* experience films. While this can also borrow from psychoanalysis in order to unpack the fantasies that viewers engage in while immersed in a film, there are other, diverse ways in which the effects, the pleasures, the sensations of watching film can be theorized, presenting a much richer picture of the potential effects of representations of science and technology.
- Last in Kuhn's nomenclature come the kinds of analysis offered by writers like Aylish Wood, which emphasize the *intertextual* relationships or networks in which films are situated. This approach, Kuhn notes, is often associated with postmodern theory.

I shall now revisit each of these approaches, and try to flesh them all out. Reflectionist analyses take film as a more-or-less transparent representation of society. Commentators have repeatedly suggested that science fiction offers particularly vivid fictionalized stories about our hopes and fears for the future, and our current preoccupations and attitudes. They portray what we are thinking, either sociologically – attitudes and social concerns – or psychologically, expressing what Kuhn calls 'the collective psyche of an era' (16). Often, this approach is periodized: it seeks to understand how cinematic representations from a particular era map on to social and social-psychological issues from that time – so that, for example, 1950s American alien-invasion films are 'really' about the Cold War, the aliens thinly disguised communists, while contemporary films depict our anxieties about genetics (e.g. *GATTACA*), artificial intelligence (e.g. *I, Robot*), or impending environmental catastrophe (e.g. *The Day After Tomorrow*).

A well-known and often-cited example of this kind of periodization and classification can be found in Andrew Tudor's (1989) study of horror movies, *Monsters and Mad Scientists*, which includes a discussion of changing representations of science and scientists in a large sample of films made between the 1930s and the mid-1980s. Although confined to the subgenre he names 'science-based horror', Tudor's analysis neatly maps a series of shifts in filmic depictions of the potential horrors brought about by science. He notes that these horrors can be either purposefully brought about by 'mad scientists' or accidentally caused by misguided uses of science. This distinction – between 'volitional' and 'accidental' impacts – is one that structures his timetable of science-horror films. So the perils of radiation or of industrial pollution are shown to represent 'accidental' scientific horrors in the 1950s and late 1970s respectively, while films from the 1930s through to the start of the 1950s more often depict mad scientists actively creating horrors, most commonly in the form of 'medical monstrosities' based on the *Frankenstein* mould.

Representations of the 'accidental' horrors caused by science, Tudor argues, underscore the idea that science is inherently a risky enterprise, and that this is the 'price' of scientific progress. But this does not always mean science inevitably produces horrors. As science is so clearly aligned with the idea of progress, it is depicted in movies as both (potential or actual) cause *and* (potential or actual) solution. This evidences the embeddedness of science in society: even once we acknowledge its potential to wreak havoc and bring forth horrors, we still keep at least some faith in science's ability to do good, to make things better. One way in which this ambivalence is worked through in cinema is by displacing responsibility. The first displacement Tudor detects is away from the individual scientist, towards science as a practice, and a practice that can corrupt individuals. Here, it is 'science' that is potentially mad, rather than the scientist – hence the emphasis on accidental horrors brought about by the quest for scientific progress, rather than the volitional horrors birthed in the labs of madmen.

The second displacement shifts blame away from science altogether: science itself becomes morally neutral, and the question of good versus bad turns to the uses to which it is put, and by whom. As Tudor notes, in films from the mid-1970s onwards, science turns bad, produces horror, in the hands of non-scientific institutions – the state, the military, or industry and business. Ultimately, Tudor (1989: 155) writes, this taxonomy of science-horror films maps on to 'a changing public evaluation of science and scientists, one that runs that gamut from fear to disillusion and disinterest'. In short, it reflects film-makers' and audiences' shared cultural understandings of and feelings about science.

The reflectionist assumption is also neatly summarized in an article on films about computers, by Anton Karl Kozlovic (2003: 345), who writes that 'Given that popular films can function as *both windows into and mirrors of society* that can impact upon public support for science, it behoves the profession to examine more closely what these pop culture narratives are telling the world' (my emphasis). His ultimate suggestion for a solution to the 'problem' of computer-phobic representation is based on an equally straightforward formula – all that is needed is 'better' representation:

> There . . . needs to be science-designed counter-educational strategies to repair the damage already inflicted upon the public's consciousness by fictional excesses . . . The scientific community could script its own feature films about possible AI worlds that are more scientifically accurate, more optimistic and not unduly focused upon neo-Ludditeism [sic], technophobia or cyberscepticism.
>
> (Kozlovic 2003: 367)

Other reflectionist critique settles on the issue of how 'realistic' the portrayal of scientific procedure is (or isn't) in films, highlighting both the possibility of using narrative cinema as a means of popularizing science, but also the dangers of 'misrepresenting' science by portraying what Lambourne (1999: 150) calls 'imaginary science', the 'deliberate hokum' used as a narrative device but which critics argue frequently misinforms audiences, damaging the task of improving the public understanding of science.

From a film theory perspective, however, reflectionism suffers from a rather different problem. As Kuhn puts it, 'the basic premises of reflectionist criticism are that the "real world" pre-exists and determines representation, and that representation portrays the real world in unmediated fashion' (Kuhn 1990: 53). However, other theoretical approaches problematize the relationship between reality and representation, including those that draw on theories of ideology associated with Marxist and post-Marxist thinkers such as Louis Althusser. Althusser's complicated ideas centre on the function of ideology to misrepresent reality to us: what we think of as real is in fact the product of ideology. So films carry ideological messages – messages about how society should work, based on the ideas of the 'dominant class'. The rest of us are subjects of and subject to ideology, but we cannot know this, because ideology is hidden away, or naturalized. We cannot see it, unless, like Althusser, we are skilled at looking for it. This is one way of understanding the narrative of a film like *The Matrix*: we are living in an imaginary world not of our own making, totally unaware of what Althusser calls the real conditions of our existence, which are fundamentally exploitative. Yet since ideology is so well-hidden, how might we detect it in something like a movie? There are a variety of tools to utilize here, including those borrowed from another theoretical tradition, semiotics, which seeks to uncover the meanings of images and stories. Importantly, ideological criticism has a political edge to it. It is concerned with disclosing the ideological work that cultural texts like films carry out – how they help sustain relations of dominance and subservience by making people think this is how things *should* be. An example of this approach is provided by John O'Neill's (1996) account of *Jurassic Park*, which sidesteps the usual focus on 'meddling with nature' and the narrative of science 'playing god' to explore instead how the film works to reproduce familial ideology – the 'naturalness' and 'rightness' of the nuclear family – through its portrayal of the relationship between Dr Alan Clark and Dr Ellie Sattler and the children Lex and Tim (also naturalized through discussions of the dinosaurs' maternal instincts and the distinction between biological reproduction and cloning in the film). Other analyses have tracked ideologies of romanticism, individualism and humanism found embedded in sci-fi movies such as *Blade Runner, Alien* and the aforementioned *The Matrix* (see, for example, Ryan and Kellner 1990; Žižek 2001).

Psychoanalytic film criticism is also about uncovering 'hidden' meanings, but here the hidden meanings are thoughts and ideas that are repressed, tucked away deep in our unconscious. These ideas and thoughts are, as Kuhn puts it, both 'unspoken and unspeakable' (91), concerning taboos and obsessions, anxieties and fixations, rooted in infanthood. Drawing on key thinkers such as Sigmund Freud and Jacques Lacan, this approach treats films like dreams, or like the 'symptoms' in patients that reveal hidden problems that cannot be addressed directly because they are so shameful:

> the primal scene [wherein a child sees its parents having sex] and the mysteries of conception and birth, sexual drives and the desire for forbidden objects, the Oedipal scenario [the child is obsessively attached to its mother and wants to displace its father as the object of her affection] and sexual difference.
>
> (Kuhn 1990: 92)

Films such as *The Terminator* and *Alien* (again – marking it as a favourite with film theorists) have been analysed from this perspective; *Alien*, in fact, has been repeatedly psychoanalysed, perhaps most notably by Barbara Creed (1990, 1993), who reads it (alongside other sci-fi and horror movies) as a working-through of unconscious anxieties about female sexuality and fecundity – male anxieties about women's reproduction projected on to the figure of the alien queen, the 'phallic mother' which men both fear and envy. Creed draws also on ideological criticism, to show how *Alien* works to uphold patriarchy through its representation of what she famously calls the 'monstrous-feminine':

> We can see [*Alien*'s] ideological project as an attempt to shore up the symbolic order by constructing the feminine as an imaginary 'other' which must be repressed and controlled in order to secure and protect the social order. Thus, the . . . film stages and re-stages a constant repudiation of the feminine figure.
>
> (Creed 1990: 140)

This theory is grounded in a series of assumptions about what goes on deep inside our heads – about unconscious processes we can never know directly, but which are thought to dominate our everyday lives. In terms of the effects that representations have, then, as Creed says, when read in this way we can see how films shore up the social order by providing a forum for those unconscious anxieties and fantasies to be addressed in a roundabout way.

By viewing the monstrous-feminine in a film like *Alien*, the audience is re-embedded in the (patriarchal) social order, having been given a cultural space to encounter its own unconscious. This leaves a large question open, however – one that is often repeated by writers hostile to psychoanalysis: how do we know what unconscious processes are occurring when someone watches a film? This troubling question leads us into the fourth of Kuhn's categories of film criticism, that centred on ideas of spectatorship. This can also be quite psycho-analytic in its approach, and falls, Kuhn writes, into two main camps: analyses based on 'enunciation' – how the film 'speaks' to its audience (here psycho-analysis explores the role of fantasy, projection, desire and so on), and the place of the viewer, or consumer, in a broader network or circuit that Kuhn calls 'the cinematic apparatus', which she describes as:

> a totalizing model of cinema as a system, a machine with many working parts: the economics of the film industry, its structures of production, distribution and exhibition, its technologies and signifying practices, its modes of reception, and the interconnections between all of these.
>
> (Kuhn 1990: 146)

Pleasure is placed at the heart of this machine, in analyses that ask questions about the experience of watching films, and in this context sci-fi films: what sensations are provoked, what feelings and moods, what thoughts do they catalyse? (See, for example, Cooper and Parker 1998; Kennedy 2000.)

One fruitful area of enquiry in this context has been science fiction cinema's use of special effects – especially in recent times CGI (computer-generated imagery) – and how this works to produce particular effects in the audience, notably the effect of 'awe' (Bukatman 1995; Cubitt 1999). This sense of awe is, in fact, two-fold: awe at the visual spectacle of the images produced, and awe at the technical feat that produced it. As Kuhn writes:

> Special effects in science fiction cinema always draw attention to them-
> selves, inviting admiration for the wizardry of the boffins and the marvels
> of a technology that translates their efforts onto the screen. They call forth
> wonder at the fictional machines of space travel, and also at the 'machine'
> of the cinematic apparatus itself.
>
> (Kuhn 1990: 148)

Hence, sci-fi films are about science and technology in a double sense, in terms of both production and representation. A big-budget movie like *Terminator 2*, for example, relentlessly exploits this twinned awe, the narrative conjuring a shape-shifting cyborg and the film-makers using sophisticated CGI technology to render a 'realistic' (but simultaneously fantastic) filmic representation. While some writers criticize some films for being overly 'effects-driven' – sacrificing plot and character to CGIs – this doubled awe is something that marks a particular kind of viewing pleasure for sci-fi audiences.[4]

Finally, Kuhn discusses an approach that lays stress on intertexuality – the way that one cultural text, such as a particular movie, relates, responds or refers to other texts (other films, books, etc.). Media commentators have drawn attention to the dense intertextuality of contemporary media culture in general, and of the centring of the audience in that intertextuality: cultural producers, such as film-makers, are seen to use references to other texts in part to 'flatter' their audience, by letting them also make the connection. This idea draws on the notion of cultural capital associated with the sociologist Pierre Bourdieu (1984), who argues that different social classes possess different amounts of different types of capital – not just money (economic capital), but also taste and style (cultural capital). These capitals are deployed by individuals and groups to mark distinctions between social class groups – you might be richer than us, but we have better taste, and so on. Cultural capital is built up through 'education' – not just formal schooling, but the on-going self-education of finding things out, keeping up with trends, building up a reservoir of knowledge. Successful media texts, whether films or adverts or novels, help to confirm our status in terms of cultural capital by making 'knowing' references or 'quotations' that only a certain 'class' of person would understand (Featherstone 1991). So a scene in *The Matrix* shows Neo hiding computer disks in a hollowed-out copy of a book by key postmodern thinker Jean Baudrillard, whose ideas (as only those 'in the know' know) are replayed in the film's narrative. Those viewers who get this 'in joke' have their status confirmed – they are, in Goldman and Papson's (1996) nice term, 'savvy spectators' – and this contributes to the critical (and commercial) success of the film.

Intertextuality is often seen to be related to postmodernity. And science fiction films are similarly often seen nowadays as commentaries on the postmodern condition, *Blade Runner* being a particularly archetypal case (Bukatman 1997). Postmodernity is attended to in Chapter 3, and is associated with the breaking down of all kinds of boundaries, including those between fact and fiction, and with a problematizing of stability and certainty. This kind of tack is at the heart of Aylish Wood's (2002) readings of contemporary technoscience films. Wood's use of the term 'technoscience' signals from the outset her ideas about mixing and complexity and about postmodern knowledge (see Chapter 3). It allows her to approach a wide range of films, not exclusively science fiction, in order to trace how postmodern knowledge or technoscience operates as a *process* rather than, in reflectionist analysis, as an object or an effect. For example, her analysis of the movie *sex, lies and videotape* steers clear of both reflectionist and psychoanalytic criticism to explore instead how the film stages technologically mediated interpersonal relationships: the video camera at the heart of the film's narrative does not *produce* alienation, but mediates between intimacy and distance, and becomes central to the renegotiation of the relationship of two central characters, Ann and Graham, precisely through its mediating function. Wood's reading of *sex, lies and videotape* reveals a network of relationships between humans and technologies which is much more complex, contingent and nuanced than any simplistic 'human good, technology bad' analysis. As Graham Thompson (2004: 98) writes, in a discussion of Douglas Coupland's novel *Microserfs*, these kinds of cultural texts 'represent the territory that is left unexamined by such a polarized debate: the way in which technology is *experienced*' (my emphasis).

One of the key issues that Wood returns to throughout *Technoscience in Contemporary American Film* is the issue of 'humanness', of how we define and redefine what makes us human, especially in relation to technology and the idea of the **posthuman**. Wood shows that the 'fictions of technoscience' she explores 'have the potential to be productive sites where new meanings of humanness as well as technologies can be generated' (Wood 2002: 184). The figuring of the difference between the human and the technological is a common trope in sci-fi cinema, of course: cyborg films like *Blade Runner* or *The Terminator* have been read, for instance, as exploring the shifting boundaries between the human and the machinic (see, for example, Pyle 2000). While Wood would want to move these analyses away from straightforward accounts of this boundary policing and binary blurring, if we follow her lead in moving outside of science fiction cinema, we can perhaps see this issue in a new light. How are these questions asked and answered in other areas of cultural production? As an illustration of what this move might accomplish, I now want to turn briefly to discuss selected ideas about and representations of science and technology in popular music.

Machine music

I want to talk about popular music at this point for a number of reasons. Like cinema, it is an important site of cultural production, transmission and consumption. It can be approached from a number of theoretical directions (see, for example, Bennett 2001; Negus 1996). It tells stories, including stories about science and technology – and these include the stories embedded in pieces of music (song lyrics, and so on) and the stories told by the 'musical apparatus', to tweak Kuhn's phrase about film: music as an industry, a commodity and an artform, produces its own responses to and representations of science and technology. As noted above, a way into this is to look at how ideas about humanness and technology have been figured in music. Here we might begin at one key node in the circuit of pop music's 'apparatus', the figure of the musician. The musician can immediately be thought of as a hybrid of human and non-human elements – the person and their instrument. However, a distinction is commonly forced here, based on an idea of artistry or musicianship: the creative aspect of music production is located firmly within the human player (or composer or producer), and the technology is seen as a passive enabler, a tool to channel the human talent of musicality.[5]

However, given the intensifying technologization of popular music production – in which key instruments such as the synthesizer, the drum machine, the sampler and the computer assume a central position – this distinction has been questioned, unsettled. In fact, as Simon Frith (1987) notes, the history of popular music is a profoundly technologized tale, even if the relationship between music and technology is sometimes figured ambivalently. And as Durant writes:

> In discussions of music, the relationship between art and technology has been an especially problematic issue during the 1980s. New modes of music-making, especially sampling and sequencing, are stimulating musical forms that exploit techniques possible only with recently available kinds of digital musical equipment; and, in doing so, the new musical technology can be thought to add to and diversify existing musical genres.
>
> (Durant 1990: 175)

This issue has been filtered through another dimension of musical performance intended to root musicality in the human: the idea of 'liveness', the symbolic centrality in terms of artistry of being able to perform 'live'. In the UK, worries over the supplanting of human musicians by technology prompted the Musicians' Union's 'Keep Music Live' campaign (Elliott 1982). Today, MTV's successful *Unplugged* series of televised acoustic concerts foregrounds this idea perfectly: idealized musicality is performed live and acoustic, unaided as far as possible by technology (but, of course, simultaneously enabled *entirely by technology*, not least the technology of broadcast; see Taylor 2001; see also Duffett 2003 on the issue of 'liveness' in relation to webcasts).

Innovations in music technology trouble such an equation of 'liveness' with musicality, leading to debates within musical circles (and their related circles, such as music journalism) about what 'live' now means in terms of musical performance: can lip-syncing be 'live', as when pop stars mime at PAs (personal appearances)? How much music can be pre-set and pre-recorded for a performance to still count as 'live'? Is programming a computer a form of music-making? (For an interesting discussion of the implications of these issues for music production and consumption, see Durant 1990.) The frame of many of these debates centres on musical cultural capital, issues of artistry and connoisseurship, and a kind of anti-technological purism that holds musicality as uniquely human. Frith discusses this in the context of rock music:

> The continuing core of rock ideology is that raw sounds are more authentic than cooked sounds. This is a paradoxical belief for a technologically sophisticated medium and rests on an old-fashioned model of direct communication – A plays to B and the less technology that lies between them the closer they are, the more honest their relationship and the fewer the opportunities for manipulation and falsehoods. [According to this position] good music is honest and sincere, bad music false and technological changes increase the opportunities for fakery.
>
> (Frith 1986: 266–7)

These issues run to the heart of what the films in Wood's book are working through: to counter this critique, in what ways might humans and technologies be brought productively together, and what's at stake in doing so?

Of course, it is not just at this level that pop music engages with science and technology; it does so at the level of content, too. Entire genres and sub-genres of pop music address science and technology themes; from industrial to folktronica and from synth pop to techno, music-makers are articulating their own responses to technoscience, positive and negative – and those which, to echo Wood once more, refuse such simple distinctions (Dery 1996; Taylor 2001; Toop 1995). If science fiction cinema represents the screening of post-modernity, then these forms of music are its soundtrack. And even beyond genres and scenes focused on technology, popular music engages with it as a theme.

One or two random examples will have to pass for sustained commentary here. First, released at the time of writing comes an album by French duo Daft Punk, known for appearing dressed as robots and for a heavily technologized (and also technostalgized) musical style based on earlier synth pop and disco traditions, and the legacy of bands such as Kraftwerk, using vocoders to 'robotize' their voices and producing an unmistakably machine-made sound. The irony comes in the album's title: *Human After All*. A band that refuses interviews and live shows, that appears as robots or animations, that produces a heavily synthetic pop, turns out to be human, after all. But what, in this context, does being human mean? Where is humanness located in this assemblage, this network? Knowingly postmodern and densely intertextual, Daft Punk exemplify Wood's (2002: 10) argument that 'cultural imaginings of technoscience . . . need not be seen as closed circuits that keep channeling the same stories. Many of them offer the potential for something different, even though the nature of that difference as yet remains uncertain'.

My second example is one I have used before, but I think it is worth returning to here: country music's engagements with technology, and the figure of the 'hi-tech redneck' (see Bell 2001). I first encountered a version of this figure when watching the Country Music Awards on television some years back. Alan Jackson came on stage, and began performing a song I later came to know as 'www.memory'. Surrounded by haybales on which stood personal computers, Jackson sang the sad lyric of lost love and hi-tech, which ends:

> You won't even have to hold me
> Or look into my eyes
> You can tell me that you love me
> Through your keyboard and wires
> No, you won't have to touch me
> Or even take my hand
> Just slide your little mouse around
> Until you see it land
>
> At www.memory
> I'll be waiting for you patiently
> If you feel the need, just click on me
> At www.memory
> If you feel like love, just click on me
> At www.memory

Now this struck me at the time, and still strikes me now, as a fascinating moment in the convergence between a certain depiction of an all-too-human stereotype, the lonesome cowboy, and the computer. Given the powerful 'frontier' metaphor used to describe cyberspace, perhaps it's not too surprising to find cowboys there (see Rheingold 1993). In fact, I soon discovered other country stars singing along the same lines: Randy Travis singing about email,[6] and George Jones's 'High Tech Redneck' ('He's bumpkin but he's plugged in'). Of course, one point these songs are making is that there's an apparent contradiction here – 'high tech' and 'redneck' seem like opposites – but in fact there's no contradiction. In a democratic and technophilic moment, rednecks are shown to be profoundly, and also *naturally*, high tech (though there are a considerable number of comic websites poking fun at this idea; see, for example, http://www.atlantaga.com/hitech.htm). As playfully, ironically and intertextually as Daft Punk, songs like these mess with ideas about technology and what it means to be human, exploring the *experience* of technology, to repeat Thompson's phrase, or the 'unexpectedly productive' comminglings of human and machine, to repeat Wood's.

There's one final angle I'd like to explore briefly here, and it returns in part to the issue of 'liveness', and its musical opposite – recording. The history of pop music is heavily imprinted with the parallel history of the development of machinery to record and reproduce music. At one end there's the recording studio, microphones, mixing desks and all that stuff. At the other, there are gramophones, wax cylinders, juke boxes, vinyl, audio tape, hi-fis, plus radios, televisions, personal stereos, and so on. And, of course, we have to add to this list more recent innovations, such as MP3 files and players. The 'popular' in popular music depends on widespread availability and circulation, enabled by technological innovation. Our ability to experience music is intensely *mediated*; it is also an experience of technology, even though the technology itself is heavily black boxed. The fuss over file sharing, while framed largely in economic terms, opened up all sorts of other questions about recorded music as a *thing* – and a thing made by and from assorted technologies (Leyshon 2001, 2003).[7]

Given the interest in **technostalgia** that runs through this book, I'd rather end by pointing to a different recorded music thing: the audio cassette. Once the mainstay of the portable music market, the audio cassette was also subject to a previous moral panic about the effects of home taping on the music industry (Frith 1987). It was also an innovation that radically transformed the practices of recording, listening to, making and distributing music (for example in the 'DIY' music subculture of the early 1980s onwards; see Rosen 1997). The audio cassette is today, however, rapidly being superseded by CD burning and MP3s, the Sony Walkman losing its market share to the Discman and then to the iPod and their countless relations (on 'redundant' music technologies, see Lehtonen 2003; see also Chapter 7). Daft Punk themselves knowingly nod to this on their home page at their record company's website (http://www. virginrecords.com/ daft_punk/index2.html), which uses an animation of a cassette tape rewinding to show the page loading – a retro image of the time spent waiting before you could listen to your music on tape (see Chapter 7). Also on the web – and a favourite of mine – is Rewind, a 'virtual museum' of blank cassettes, with countless scanned images of tapes and their cases (http:// www.studio2.freeserve.co.uk) lovingly archived for future generations who never knew the joys (and the trials) of home taping.[8]

So, even when barely glimpsed through these scanty examples, we can hopefully begin to think about the many ways popular music has engaged with science and technology, in terms of genre, theme and form. Added to my lengthier discussion of film, these two major domains of popular cultural production and consumption can be seen to contain a multitude of ways of telling stories about science and technology. While the public understanding of science approach stresses the straightforward transparency of representation – so the 'hi-tech redneck' would be seen as a positive representation of technology (and I guess also of rednecks) – I'd prefer it if we told more *complicated* stories. Not complicated for the sake of complication, but complicated in recognition that the interactions and interfaces between science and society, or culture and technology, or however you want to frame things are, to use Megan Stern's (2003: 164) terms, 'polymorphous and pervasive', informing our lives 'in multiple, complex and intimate ways'.

Further reading

Allan, S. (2002) *Media, Risk and Science*. Buckingham: Open University Press.

Haynes, R. (1994) *From Faust to Strangelove: Representations of the Scientists in Western Literature*. Baltimore MD: Johns Hopkins University Press.

Kuhn, A. (1990) (ed.) *Alien Zone: Cultural Theory and Science Fiction Cinema*. London: Verso.

Taylor, T. (2001) *Strange Sounds: Music, Technology and Culture*. London: Routledge.

Tudor, A. (1989) *Monsters and Mad Scientists: A Cultural History of the Horror Movie*. Oxford: Blackwell.

Turney, J. (1998) *Frankenstein's Footsteps: Science, Genetics and Popular Culture*. New Haven CT: Yale University Press.

Wood, A. (2002) *Technoscience in Contemporary American Film: Beyond Science Fiction*. Manchester: Manchester University Press.

Notes

1 I am aware I'm treading a very dangerous line here, in suggesting that cinema-goers are 'non-specialists'. Certainly, it is not my intention to suggest that popular forms of expertise are any less important or valuable than academic ones. Rather, I am trying to make the point that 'common-sense' readings of films do matter – in fact, that they are the most significant for a whole host of reasons. It is easy for academics to lose sight of this as we happily articulate dense theoretical positions and offer complex readings of films drawing on our favourite bodies of theory. See Irwin and Michael (2003) for a similar point about the separation between scientists and laypeople – a separation that oversimplifies the borders of scientific knowledge, but nevertheless a separation that in itself produces important effects.

2 I have been struck, when trying to exemplify this typology for students, that Hollywood actor Jeff Goldblum has made a career out of working through these stereotypes, through his roles in movies such as *The Fly, Jurassic Park, Independence Day* and *The Life Aquatic with Steve Zissou*.

3 My twist on the title of a famous edited collection, *Film Theory Goes to the Movies* (Collins *et al.* 1993).

4 Of course, CGIs are not solely used in sci-fi film-making, as effects-heavy block-busters such as *Gladiator* and *Titanic* remind us; nevertheless, the closeness of the connection, the ways that viewing attention is captured by special effects, and the relationship between effects and narrative, mark the relationship of CGI and sci-fi as particularly intimate (see Bell 2001).

5 For an interesting and provocative discussion of what musicality means, which resonates with my discussion, see http://www.hornplayer.net/archive/a212.html

6 Does anyone know what this song is, or if it even exists? Maybe I imagined it, or maybe it's not Randy. Email me if you know.

7 Of course, at the same time, MP3 file technology *dematerializes* recorded music, turning it into code and stripping it of its associated materialities.

8 All websites in this chapter were accessed in March 2005.

5 | THE MOON AND THE BOMB

I grew up in space.

(Constance Penley)

In this chapter, I want to think through much of what has come before through the lens of *experience*, by looking at two very powerful symbols of science, technology and culture. The chapter is about living in a technoscientific culture, and it's about me. Born in the Midlands of England in the mid-1960s, and growing up through the 1970s and 1980s with my father employed as a high-school science teacher, my youthful engagements with science and technology were overshadowed by two emblems: going to the Moon, and the threat of nuclear war. The **Apollo** Moon landings, and their promise of a future of Moon bases, space stations and interplanetary travel, represented the utopian possibilities of the age, and of ages to come. Daydreaming about the year 2000, hard as it was to picture being 35 years old, conjured images blended from sci-fi shows and movies, and populated by machines I made kit models of. (Interestingly, the other future I routinely imagined – and drew in intricate pictures – was life at the bottom of the oceans, inspired at least in part by my fascination for *The Undersea World of Jacques Cousteau*.) This was the version of techno-scientific progress my father also fostered, with his faith in science (and in science pedagogy) as a window on to the truth of the world, the universe, and most other things, and with his enthusiastic embrace of new (including nuclear) technologies – again, suggested by our regular viewing of shows like the BBC's long-running and immensely popular *Tomorrow's World*, which showcased inventions and innovations.

By contrast, the idea of nuclear war existed in the background, as an ambient fear. It was rarely discussed explicitly in front of us kids, and through a child's eyes, events in the Cold War were easily rewritten as a matter of technological prowess and progress rather than diplomacy and brinkmanship. Unaware of the geopolitical backdrop, there was nevertheless a pervasive low-level (yet rarely articulated) sense of threat, of fear, especially as we lived near two major industrial sites – one a chemical plant, the other making aircraft engines. (The possibility of these factories becoming bomb targets was reinforced by the continued presence of derelict World War II gun emplacements on farmland near my house – a significant site for childhood games, but also a powerful reminder of past and potential vulnerability; see also Virilio 1994; Woodward 2004.) The ambient nuclear fear was experienced more commonly as the subject of playground chatter, with half-overheard rumours about three-minute warnings and the installation of sirens, but we were too young to really feel the more pervasive adult anxieties of the period, so vividly described by Martin Amis (1987). (We were, of course, shielded from these anxieties by our parents, by teachers – we were never conscripted into practice shelter-building or anything; though we did talk and fantasize about what we might do with our 'last three minutes' once the sirens went off.) This was our 'nuclear normality' (Lifton and Markusen 1990), the banality of living with the Bomb as an abstract backdrop to the relations of everyday life. While not as blasé as the 'atomic apathy' Margot Henrikson (1997: 194) describes in early 1960s America, this backgrounding of anxiety represents a similar coping strategy for living under the on-going mental fallout of nuclear threat.

It is interesting that my boyhood understanding and appreciation of (even enjoyment of and obsession with) war more generally was in fact *all about technology*. World War II was repackaged for youngsters like me as a series of technological marvels. War machines were made available to us, whether in Airfix kit models, or Action Man and his ever-growing battalions and arsenals, or weapons-obsessed partworks and Top Trumps cards. Again, we only dimly understood the grim realities of the War, and its political contexts (a fact that must, in retrospect, have been at times difficult to handle by our parents, who had lived through the War – a fact only brought home to me when my uncle, who had fought in the 'Desert Rats', asked why I had a Nazi poster on my wall!). We were interested not in the politics of the War, but the paraphernalia, the hardware. This disposition towards war – a routine element of being a boy at that time, supported by endless commodities and media texts – to some extent spilled over into my equally unpolitical understanding of nuclear war. Recent TV programmes like the BBC's *Crafty Tricks of War* replay this disposition, with war games shown as feats of technological ingenuity – albeit often in a very British, amateur, tinkering kind of way (a point I shall return to later). The World War II media feeds I was given parental consent to access – such as the long-running BBC drama series *Secret Army*, about the French resistance movement, or tales of POW escapees from Colditz – equally focused on the crafty tricks of espionage rather than the bureaucratic rationalities of Nazism (Bauman 1991).

So dim was my understanding of the Cold War that I blurred it in my mind with the less globally-threatening incident over North Sea fishing rights, known as the 'Cod War'. Only as I grew older, and began to enter into political (counter) culture, did I come to see the full import of the shadow I'd grown up under – and the rather perverse experience of boys' war games. Until then, I had preferred to fix my childhood gaze on the scale model Apollo Saturn V rocket in my bedroom, oddly poised for launch amid an assortment of World War I and II aircraft. (I'm pretty sure Airfix didn't make equivalent scale models of ICBMs or other parts of the nuclear armoury; and, of course, the kinds of media texts to which I was granted access never spelled out the connections between the arms race and the space race.)

This ambivalent pair of Cold War twins, the Moon and the Bomb, is beauti-fully captured in two books by the American photographer Michael Light – two publications that have become my favourite coffee table books, for their ability to capture the wonders and the horrors of the age. In *Full Moon* (1999), Light has collected photographs from NASA's moon missions, while in *100 Suns* (2003) he collates images of US nuclear tests. Page after page of incredible photos, truly sublime images, at once familiar and amazing: rockets, moon-scapes, beaming astronauts; mushroom clouds, deserts and atolls, bedazzled servicemen. The place names are equally exotic and evocative, especially the iconic pairing of The Sea of Tranquillity and Bikini Atoll (for a short essay on Light's work, see G. Thompson 2004).

But the public history of these images and their consumption reveals other forces at work. The photographic records of the space age and the atomic age produced particular (intended and unintended) effects for their viewers. On the one hand, as Scott Kirsch (1997) argues, publicly circulated images of atom bomb tests in the US spectacularized the mushroom cloud as technological and aesthetic marvel while simultaneously obscuring the toxic fallout of the tests, making bomb photos a powerful part of the pro-nuclear PR war. On the other hand, the photographs from space – particularly images of Earth shot by astronauts – stirred a nascent anti-nuclear sentiment: 'The ability to visualize the Earth in a single image coincided with the presumed ability to destroy the Earth with one massive unleashing of weapons . . . [B]oth accomplishments were byproducts of the same endeavour' (Vanderbilt 2002: 50). This new imaging of the Earth as both beautiful and vulnerable became a stock icon for the environmental movement, reproduced on protest posters and placards as well as becoming the powerful cover image of the *Whole Earth Catalog*, a 'countercultural consumer publication that combined a freely adapted, holistic ecological scientism with a practical set of lifestyle injunctions and techniques meant to bring about both social and personal renewal through a practice of environmental global stewardship' (Binkley 2003: 287).

There is a powerful resonance between the two sets of images, then, as both stand as records of a time when science could do pretty much anything, and the things it did were to be marvelled at – just as NASA launches drew large and enthusiastic crowds, hotels in Las Vegas (and doubtless elsewhere) hosted themed bomb parties to give guests prime views of Nevadan atomic tests (Lang 1952/1995). Consider these two descriptions:

> During the 1950s and 1960s, while the consciousness of many Americans was haunted by the threat of nuclear annihilation, some [Las Vegas] boosters promoted the explosion of nuclear devices outside of town as an entertainment event . . . [O]ne enterprising promoter, Desert Inn owner Wilbur Clark, timed the gala opening of his new casino, the Colonial Inn, with that of an atomic bomb blast in March 1953 . . . One [publicity shot] showed a girl sporting an Atomic Hairdo, the product of a Las Vegas beauty parlor. Another heralded the Atomic Cocktail, invented by a local bartender in one of the hotels.
>
> (Gottdeiner *et al*. 1999: 79)

> The *New York Times* reported [in 1969] that Apollo XI attracted 'the largest crowd ever to witness a space launching', amounting to more than a million and including 'former President Lyndon B. Johnson, members of the Poor People's Campaign, African and Asian diplomats, youth carrying Confederate flags, vacationing families, hippies, scientists and surfers, and students and salesmen'.
>
> (Nye 1994: 241)

As Nye suggests, over time the public fascination with nuclear testing waned, with secrecy rather than publicity coming to determine the structure of these events – and, of course, following the Limited Test Ban Treaty of 1963 the bombs retreated underground, the image of the mushroom cloud turning 'from icon to artifact' under these new proving conditions (Kirsch 1997: 246). At the same time, the continued desire to spectacularize NASA launches – with the Agency having to endure the public's fading attention span (Penley 1997) – can be read as a move to overshadow the nuclear arms race with the space race, to rewrite the Cold War in terms of 'progressive' technological endeavour (going into space) rather than mutually assured destruction:

> Whereas a space launch awakened the will to believe, the bomb evoked uncertainty and dread . . . Public enthusiasm for the space program represents a nostalgic return to the technological sublime, a turning away from the abyss of the nuclear holocaust . . . The affirmative sense of

achievement that followed the moon landing made government high-tech programs attractive at the very moment when the nuclear stockpile was large enough to destroy every living thing on earth.

(Nye 1994: 254–6)

In this sense, the powerful utopian dreaming of space missions is the benign twin of nuclear destruction; an attempt to harness the terrifying power of technology and 'rescue' it from an overdetermined association with death. The space race therefore becomes an adjunct of '**nuclearism**' – the shaping of discourses about nuclear technology away from their catastrophic connections, most commonly through the euphemism of 'deterrence' (Irwin *et al*. 2000). Images of space and images of the Bomb commingle in what Matthew Farish (2003: 132) labels the 'popular geopolitics' of the Cold War period.

Of course, there was a second 'diversion' offered to the public, to avert their eyes from the Bomb's light: the 'peaceful atom', nuclear power, with its futurological promise of 'Nuketopia' (Wills 2003: 153). However, as Ian Welsh clearly delineates, this proved a difficult ideological manoeuvre. The 'nuclear bargain' that promised energy (and with it prosperity, abundance, etc.) could not do the necessary cultural work to conscript the public into a nuclear future, to separate 'welfare' from 'warfare' (Irwin *et al*. 2000). Nuclear power would, it seems, be forever ghosted by mushroom clouds, no matter what attempts were made to domesticate it. When my sister returned home from college with a 'Nuclear Power No Thanks' bumper sticker on her VW Beetle, her 'irrationality' incensed my father, whose scientistic mindset was able to do the work of hiding the mushroom cloud behind the peaceful atom. (As Amis (1987: 16) writes, on the nuclear issue 'we are all arguing with our fathers'.) Welsh notes that the desire of physicists for a nuclear future that promised betterment for all humankind was a powerful discourse, to be sure. Nevertheless, among the non-scientific and non-military-political, 'belief and recruitment were never wholehearted' (Irwin *et al*. 2000: 99), just as other attempts to find 'peaceful' uses for nuclear technology, such as earthmoving by atomic explosion, could not decouple the bomb from death (Kirsch 2000). In an age where the new euphemism of 'WMDs' has come to stand in for MAD, and where possession of a nuclear arsenal is cast as an act of aggression rather than defence or deterrence, the chances of recruiting to the side of the peaceful atom seem slimmer than ever. Perhaps this also helps explain George W. Bush's reheating of the US space programme – possession of nukes no longer instils national pride and feelings of security; it becomes even more pressing to obscure the mushroom clouds with the benevolent, other-worldly and futuristic technologies of space exploration.

To infinity – and beyond

Bush is doing much more than tapping the *zeitgeist*, of course. As Constance Penley powerfully and persuasively argues in *NASA/TREK* (1997), ' "going into space" – both the actuality of it and its science fiction realization – has become the prime metaphor through which we try to make sense of the world of science and technology and imagine a place for ourselves in it' (Penley 1997: 5). In this section I want to explore this contention in some detail, and to work through the salient points of Penley's analysis of the role of space exploration in the American imaginary. Starting with the premise that 'NASA has by now become popular culture' (3), Penley charts the 'popularizing' of the American space programme, and its interweaving with other pop-cultural space programmes, especially *Star Trek* – hence the twinning in her title. Her deeper question concerns public understanding of science, and science's understanding of its public. *NASA/TREK* gives her an entry into issues of the status and roles of 'popular science' in America, and of the 'going into space' metaphor and project as central to those issues. Defending 'popular cultures of science' and the work of popular culture in helping people think (and dream) about science and technology, Penley works to (re-)define 'popular science' as an on-going creative engagement with both the 'facts' and 'fictions' of science:

> Popular science involves the efforts of scientists, science writers, and scientific institutions to attract interest and support for advancing science and technology. Popular science includes many science fiction television shows (and fewer films) that offer a personalized, utopian reflection on men and women in space. Popular science is fictional work that carries on this reflection. Popular science is ordinary people's extraordinary will to engage with the world of science and technology. Popular science wants us to go into space but keep our feet on the ground. Popular science envisions a science that boldly goes where no one has gone before but remains answerable to human needs and social desires. Popular science, fully in the American utopian tradition . . . insists that we are, or should be, popular scientists one and all.
>
> (Penley 1997: 9–10)

In this project, Penley charts the uses and meanings of NASA in popular culture, from filmic portrayals to folk humour in the wake of the *Challenger* disaster, and from the Martian microbes to the on-going problem of maintaining public interest in an increasingly repetitive (and bungle-prone) programme of missions.

Set alongside this is a second popular science narrative: *Star Trek*. As a rare utopian vision of the future, *Star Trek* is interwoven with NASA in dense and complex ways, condensed by Penley into the formula: '*Star Trek* is the theory, NASA the practice' (19). *Star Trek* is shown, like NASA, to inhabit a multitude of symbolic cultural sites, to spill out from the TV screen into pop-science books and into everyday parlance ('Beam me up, Scotty', 'It's life, Jim, but not as we know it' and of course, 'To boldly go . . .'). But what interests Penley more is the merging of the two, the dense intertextuality and cross-referencing that produces her hybrid *NASA/TREK*. In particular, this intertexting is read by Penley as NASA attempting to sustain popularity by appealing to *Star Trek*-like narratives – to re-enchant the space programme in the face of both its increasing routinization or mundanization (especially in the Shuttle era) and its spectacular failures and disasters, particularly *Challenger*. Penley argues that *Challenger*'s **Teacher in Space Program**, a massive PR campaign to put the Shuttle missions back in popular favour, produced exactly the opposite effect. Moreover, the explosion of the shuttle soon after take-off, screened nation-wide (and beyond) as part of the edutainment package, profoundly rewrote the script of 'popular science' in the US: 'NASA's attempt to make the space program popular with young people all across America literally blew up in its face' (Penley 1997: 41). When Penley quizzed her own students about their experience and recollections of the disaster, many stated that it was at that moment that they became disenchanted with science.[1]

Unpacking the meanings of *Challenger* is central to *NASA/TREK*, but I want to focus on the broader issue here. That is NASA's need to repeatedly re-enchant space science, in order to continue to legitimize the enormous budgets allocated to its missions. In forever needing to be positively newsworthy, utopian and aspirational, NASA has had to be uncommonly *zeitgeisty* for a large sci-tech organization. Arguably it has maintained its public profile not only through its *Star Trek* makeover, but also by bending its mission to tap the popular imagination. And the popular imagination is notoriously fickle and channel-hoppy. So, while Gary Edgerton (2004: 136) is spot on in calling the televised NASA missions 'an early and unusually compelling brand of reality programming', like all reality TV shows, its viewers began to drift away, distracted by other matters.

In the mid-1990s, no doubt mindful of the dwindling audiences for its endeavours, NASA switched its emphasis from the democratic dream of mass space transit (embodied, unsuccessfully, in the Shuttle), and sided with growing public interest in UFOs and ETs (Dean 1998).[2] NASA's role became the search for life in space, rather than the attempt to put human life up there, and the Martian microbe saga stands centre-stage in the rebranding of NASA as the 'real' *X-Files*: 'No expectation has had more influence on support for space

exploration than the belief that humans are not alone' (McCurdy 1997: 110) –
and that *the truth is out there*. The enormous worldwide media circus that
centred on one small piece of space rock, labelled ALH84001, promised, at
least for a moment, to provide exactly the re-enchantment that NASA badly
needed:

> Judging from the crush of journalists, one would think that a movie star or
> perhaps the president had entered the room. Photographers gathered
> around the subject, taking pictures from every conceivable angle, their
> flash guns popping over and over, while television cameras zoomed in,
> searching for a view that was up close and personal. But the object of this
> attention was no human celebrity. Rather, it was a piece of rock in a clear
> case. The rock was a piece of Mars that had fallen to Earth as a meteorite.
> Moments earlier, a team of researchers had told a packed press conference
> at the Washington headquarters of . . . NASA, and a worldwide television
> audience that watched it live, that they had found evidence of fossilized
> Martian bacteria in the rock.
>
> (Kiernan 2000: 15)

While the warm glow of the media spotlight was short-lived, and soon overcast
by doubts over the veracity of the claims made by NASA and by their handling
of the flows of information to the press and thereby the public, the cultural
effect of this moment resonated for much longer.

At a time of intense interest in 'life out there', buoyed up by endless pop-
cultural representations, the possibility of bacterial life on the Red Planet was
a source of great popular excitement. In part this taps the longer-running
construction of Mars as inhabited by space aliens, which reaches back to
Schiaparelli's description of '*canali*' on the planet surface in the nineteenth
century – transliterated into English as 'canals' (McCurdy 1997). It also taps the
parallel vein of sci-fi writing and later screening of life on Mars, the site of both
fascination and terror. So, while further research into the meteorite contested
the original findings, concluding it did not provide convincing proof of past life
on Mars (Kiernan 2000), ALH84001's role in reshaping 'popular science' in the
Penley sense has been paramount. It has fed into and also fed off the swirl of
popular interest in all things extra-terrestrial, and also undoubtedly worked to
counter the **conspiracy theories** of government cover-ups and secrecy surrounding alien life (unless, of course, your branch of conspiracy theory sees
ALH84001 as little more than a smokescreen to divert attention from the 'real'
discovery of alien life, whether in Roswell, New Mexico or any other of the
many secret locations on the conspiracy buff's atlas – see Chapter 6; Dean
1998; Paradis 2002). So, whatever the 'truth' of the Martian meteorite and its
microbes, it helped fulfil that other key *X-Files* mantra: *I want to believe*. It also

folds into the intensifying focus of space agencies on visiting Mars to search for 'signs of life' – or at least signs of the past potential conditions for life, especially in the form of water. Mars is currently being orbited and roved over by a number of devices, the latest in the 30-plus attempts to get close to or on to the planet. At the time of writing, Mars is very much in the news, as it has been here since the development and subsequent disappointing disappearance of *Beagle 2*, Britain's contribution to the joint NASA-ESA Mars Express programme (see later).

The search for 'life' in space is, obviously, a subject that captures very vividly the popular imagination, played out across popular culture and becoming central to 'popular science'. One of the most interesting and novel aspects of the popular science of life in space is the SETI@home project. The broader troubled history of the **Search for Extraterrestrial Intelligence** (**SETI**), especially its relation to NASA and US Congress, need not detain us too long here (see Squeri 2004). What is worth noting, however, is the birthing of SETI in the context of nuclear threat, with part of the promise of finding alien life being tied to the hope of pedagogic revelations from ET about either how to avoid annihilation, or how to survive it:

> When the possibility loomed of thermonuclear warfare destroying humanity in several minutes, many Americans were gripped by a deep anxiety. One response to this angst was to believe in the existence of extraterrestrial civilizations that had survived the invention of lethal technologies. These alien civilizations would be older than Earth's, more advanced, and peaceful. If humans could somehow contact these advance civilizations, they might learn the secrets of survival.
>
> (Squeri 2004: 478)

But how could contact be made? The SETI approach is to listen out for radio signals, filtering out background noise. As Marina Benjamin (2003: 184) notes, this idea is premised on a very narrow and peculiar way of thinking about life 'out there': 'researchers listen for radio signals on the assumption that alien civilisations are either deliberately or unintentionally broadcasting'. As the Earth currently leaks radio waves constantly, the SETI rationale is that somewhere, someone or something might have developed the same technology and be doing the same thing. But, as a SETI scientist conceded to Benjamin, this is a very, very small window of probability, considering the minute slice of astronomical time that we have been using broadcast technologies here on Earth: 'For the first four billion years of life on Earth we didn't leak radio at all, then suddenly a hundred years or so we leak like crazy, and now, if we go digital, we will return to being radio-quiet' (Benjamin 2003: 186). Further, other civilizations might not be using a form of communications technology we can detect, or even dream of: 'it is quite possible that ET may be phoning but that we don't know how to listen' (Squeri 2004: 487).

The other key factor inhibiting the chance of SETI being successful is the opposite of this impossibly narrow time-window: the vastness of space itself, and the enormous amount of computational power needed to process all the data collected by the telescopes, to sort out the noise from the (potential) signal. And that is where SETI@home comes in. Benjamin succinctly summarizes this system thus:

> It is an ingenious free-subscriber program that enables a California-based team of astronomers, physicists and engineers to borrow computing power from ordinary people eager to participate in a bona-fide scientific experiment, and then to put that processing power to use in analysing radio signals from outer space for signs of intelligent life. Anyone can climb on board the enterprise, anywhere in the world, and help search for electromagnetic evidence that we are not alone in the universe. All that is required is a reliable PC with idle processing time to loan out, and access to the Internet.
>
> (Benjamin 2003: 168–9)

SETI@home offers a very tidy and also very appealing solution to the problem of crunching all that data, therefore. It conscripts 'ordinary people' and their PCs into helping (albeit passively), exploiting at once the spare capacity lying idle in most folks' underused computers, the ease of the Internet to transmit data worldwide, and also the popular appeal of SETI's project. In return, participants are able to watch, via a screensaver, the data being processed by their PC, and enjoy the serious fun of being part of the search. It has proven incredibly popular, with 3000 new users per day downloading the program, and participants by 2002 numbering more that 3.5 million in over 200 countries. It represents a new version of one of the original drivers behind the development of the Internet – time sharing (allowing remote users access to idle processing time; see Abbate 2000), repurposing it into, as Benjamin says, 'the largest community computing project that the Internet world has ever seen . . . SETI@home has in short order become the world's most powerful supercomputer' (172, 181). It has also spawned its own 'subcultures', with devotees chatting and sharing more than just processing power, as well as providing a model for other large-scale data-processing efforts. The way in which SETI@home has tapped that desire for 'popular science' that Constance Penley points up in American culture is truly remarkable. Even the aesthetics of its interface, with a kind of *Star Trek*-like retrofuturism, make users feel as if their computer is doing the things we imagined computers would be doing, rather than the humdrum business of typing, filing and storing. It is, in short, probably the best representation of 'popular science' today.

SETI@home is, on the evidence of participants, 'fan' sites and so on, also a truly global representation of popular science. The question of the global reach of the popular culture of science was always one that nagged at me in my readings of Penley's *NASA/TREK*. Where she rightly states that 'science is popular in America' (4), and argues persuasively for the role of space travel in making and marking that popularity, I always wondered how cross-culturally transferable her claims were. When teaching *NASA/TREK*, I always ask my students to ponder this, too, and to do the imaginative work of conjuring their own childhood dreams of science and technology, and of space. And, while globalized media flows are argued to have Americanized us all, the students often reported a different, distinctively un-American attitude, both in terms of their readers' responses to texts like *Star Trek*, and in terms of the indigenous pop-cultural products they were and are exposed to, which had a different aesthetic and a different sensibility. In place of *Star Trek*'s utopianism, my students talked nostalgically about TV shows like *Doctor Who, Blake's 7, The Clangers* and *Button Moon*. So, with heartfelt thanks to them for the many illuminating classes spent unpacking these and other shows and movies – and tweaking the title of Howard McCurdy's excellent book – I'd like to move on now to think about this issue.

Space and the British imagination

In the introduction to *British Science Fiction Cinema* (1999: 1), I.Q. Hunter attempts to define the peculiar Britishness of these films, with their 'specific, often quirky inflections of the genre'. Hunter surveys post-war sci-fi movies, noting that while they share common themes with their US counterparts, in terms of Cold War motifs, anticommunism and nuclear paranoia, there are other equally powerful narratives at work, in particular the collective working-through of World War II and especially London's Blitz – seen, for instance, in the *Quatermass* films. There's also a set of British sensibilities depicted in terms of both the state's and ordinary people's responses to alien invaders. As Hunter writes, in these pictures, 'aliens turn up in the most out-of-the-way places, and the action is often set in country pubs, cheap locations where a cast of British stereotypes just sit around, talk and react with various degrees of sang-froid to the disruption of their quiet lives by symbols of modernity and Otherness' (8–9). The 'threat' of modernity, in terms of the social upheavals of the post-war period, also mark British sci-fi cinema in particular ways, trading again nostalgically on the wartime cohesion of the 'spirit of the Blitz'. More recent films, such as *Brazil, When the Wind Blows* and *The Hitchhiker's Guide to the Galaxy* continue this tradition in terms of picking up British themes and sci-fi-zing them within a recognizably British *mis-en-scène*.

Like the films that Hunter and his contributors discuss, British sci-fi TV similarly displays its own approach. In *Button Moon*, a wooden spoon travels into space in a baked bean tin; in *The Clangers*, the Moon is populated by knitted aliens. Both series have a hand-made, low-tech aesthetic, equally at work in *Doctor Who* or *Blake's 7* (though the latter might have attempted to be more glossy, more American, its low budgets and use of home locations never lent it *Star Trek*'s veneer). Perhaps the contrast is most stark, in fact, if we look at *Doctor Who* and *Star Trek* side by side. In place of the Enterprise, for example, the eponymous Doctor had his Tardis – a space/time ship inside a police telephone box (though admittedly the Tardis was much more of a technological marvel than the Enterprise in some ways, being able to travel through time, and being bigger on the inside than the outside). As Nicholas Cull (2001: 99) writes, *Doctor Who* and the Tardis signify Britain's exclusion from the space race: 'Only Americans or Russians . . . got to go into space. The *Doctor Who* format enabled the British to imagine themselves in space by leaping over the problem of the spacecraft. The Tardis required neither expensive special effects nor a complex plot to imagine a British space programme' – in fact, a *very* British space programme. Moreover, Cull argues, the show was more focused on working through World War II than the Cold War. While it did deal with Cold War issues, such as the threat of nuclear annihilation – most consistently through the depiction of the Daleks as post-atomic mutants – it nevertheless suggested a World War II response, harking back to Britain's wartime stoicism:

> Britain's experience in the Second World War, projected forward into the nuclear age, becomes the ideological touchstone – in particular, the Churchillian romance of the plucky underdogs at their best only when called to fight; confronting impossible odds for the ultimate 'rightness' of a cherished democratic principle. [*Doctor Who*] contemplates the scenarios that might attend the onset of a Third World War, yet it can only conceptualise these according to national myths surrounding Britain's experience in the Second.
>
> (Cook 1999: 125)

The characterization of the Doctor himself, through all his various 'regenerations', perfectly encapsulates the Britishness of the show, trading on an assortment of well-defined national archetypes (see also Tulloch and Jenkins 1995):

The characters of the Doctor incorporated a number of British types including H.G. Wells' time traveller; Sherlock Holmes; Van Helsing in *Dracula*; and the ever resourceful back room boffins who according to British popular culture had won World War Two by inventing bouncing bombs. The boffin had become part of British science fiction in the 1950s . . . Moreover, the Doctor's adventures repeatedly showed the triumph of brains over brawn, a national trope that could be found everywhere from school fiction to prisoner of war drama.

(Cull 2001: 100)

While Cull notes that the series' version of Britishness became increasingly old-fashioned and conservative, contributing to its ratings decline and eventual demise (though, at the time of writing, the show has been reprised, with the BBC giving it a contemporary take to attract today's viewers),[3] there are threads of this 'British space programme' picked up in later films and shows, including the animated adventures of Wallace and Gromit, *A Grand Day Out* and *The Wrong Trousers*. In the latter, for instance, hobbyist inventor Wallace buys a pair of cast-off robotic trousers from NASA, which he fashions into a dog-walking machine (the trousers are subsequently appropriated by the pair's lodger, a mysterious and devious penguin, who uses their gravity-defying and remote-operating capabilities in a daring diamond robbery – a fine example of the double life of technology discussed in Chapter 3).

As Cull notes in relation to the character(s) of the Doctor, and as Wallace equally illustrates, the figure of the boffin is iconic in British popular science – the hobby scientist, the tinkerer, working with limited resources but unlimited imagination (see Chapter 4). This figure is hymned in Francis Spufford's (2003) *Backroom Boys*, a book which also has some valuable comments to make about Britain in space. Spufford notes the apparent incommensurability of the idea of space travel with the ambitions of British post-war science (and politics). By the mid-1960s, he writes, 'the naïve dream of Britain in space had become a ghost, a shadow' (23), while by the 1980s, Britain (and especially its government) was 'allergic to rockets' (5). He does acknowledge, however, that Britain did have a very modest space programme from the 1950s to the 1970s, vividly evoking the scene of rocket science in Cumbria: 'Men in tweed jackets with leather elbow patches sat at control rooms watching bakelite consoles. The countdown was heard in regional accents' (ibid.). However, while Britain always lagged behind in the race for space, in another area of future-facing technoscience it participated to produce a true marvel: Concorde, which Spufford rightly casts as Europe's equivalent of the Apollo space programme, a plane so sci-fi-tastic that it 'still looks as if a crack has opened in the fabric of the universe and a message from tomorrow has been poked through' (39) – hence the lachrymose, end-of-an-era nostalgia that marked the plane's recent decommissioning. And in terms of British space missions, the final chapter of *Backroom Boys* previews and backstories the visit by yet another peculiarly British contraption, *Beagle 2*, to the distant deserts of Mars:

> The British *Beagle 2* mission is the most wonderful, zany, off-the-wall, left-field, ad-hoc, put-it-together-in-the-spare-room, skin-of-its-teeth project of my or anybody else's lifetime. . . . [A] bunch of harebrained British scientists, engineers, techno-freaks, geeks, nerds and saddos [has] decided to build a little dog, hitch a lift on a rocket, and fling it across the void to Mars. And all without a decent pullover.
>
> (Appleyard 2003: 32, 38)

The (human) public face of *Beagle 2*, who was omnipresent across British news media around Christmas 2003, was Colin Pillinger, head of the Planetary and Space Science Research Institute at the Open University.

Pillinger is every bit the British boffin, a fact he undoubtedly played up to the press, with his endless use of soccer metaphors to account for Beagle's progress (as well as to temper the ultimate disappointment of its vanishing on the planet's surface). Spufford sketches him like this:

> Pillinger [has] huge mutton-chop whiskers and a strong West Country accent that [makes] him sound like a man perpetually leaning on the gate of science with a straw in his mouth. At weekends, he [likes] to relax with his herd of dairy cows.
>
> (Spufford 2003: 216)[4]

In fact, the media's 'Beagle-talk' endlessly played on the boffin or 'Spitfire Man' motif, and the British triumph-over-adversity-by-ingenuity trope, with its implicit critique of the big boys of technoscience, such as NASA. Here was a vivid example of British pluck, self-deprecating but simultaneously self-aggrandizing: '*Beagle* owned up to being a little bit comical, which let it aspire to being a little bit brave as well' (Spufford 2003: 224). Pillinger also revealed himself to be sufficiently media-savvy to whip up what Appleyard labels 'Beaglemania' (36) in the national imaginary, enrolling some key figures associated with New Labour's 'Cool Britannia' project (such as Britart's Damien Hirst and Britpop's Alex James), to tickle our collective cultural fancy. And while it is named grandly after Darwin's discovery vessel, *Beagle 2* also conjures other images, of the heroic but vivisected space dogs from the early years of rocket science (Hankins 2004), as well as Britain's more recent imagined canine astronauts, notably *Doctor Who*'s robodog companion, K9, and the animated mutt Gromit, who visited the Moon with his owner Wallace in the short film *A Grand Day Out*. (Echoing intertextually the narrative of that film, where the heroes encounter a weird, alien, robotic oven on the Moon, Spufford describes national pride in Britain being 'the country that put a barbecue on Mars' (224).)[5] A nation of dog lovers couldn't help but love *Beagle 2*, and putting a garden-shed made, low budget, sketched-on-a-beer-mat robot dog on Mars therefore makes the mission another episode in the long-running 'British space programme', in both material and symbolic senses.

At the time of writing, *Beagle 2* has yet to send its call signal, its Britpop woof, back to Earth. No one as yet knows exactly what's happened, just that things haven't gone 100 per cent to plan. While all hope is not lost – never losing hope being very much a part of the British way with these things – at present our little space dog is nothing more than another piece of crash debris on the surface of Mars, part of the ever-growing litter left behind by space missions successful and unsuccessful. The flotsam and jetsam of the space age, all the left-behind, used-up, gone-wrong bits and pieces, are a strange legacy of this scientific endeavour, in some ways a typically human, throwaway symbol of the cost and the waste of space programmes. While some artifacts of the space age are lovingly (if complexly) preserved for future generations to goggle at (e.g. spacesuits – see Lantry 2001), there is something almost melancholic about the abandoned, marooned, broken debris that surrounds us in space. We might imagine these cast-offs being picked over by future astro-archaeologists – or even ET-archaeologists, like those depicted in the movie *AI*, trying to piece together a story about the Earth's past (Bell 2004a; Rathje 1999). This kind of archiving of the present as the past of the future can be linked to a broader, ambivalent project that I now want to turn to: the 'museumifying' or 'heritizing' of our recent technoscientific past, and the issue of dealing with the relics, remnants and remainders of the space age and the nuclear age, both intended and unintended.

Downwind

To bring this chapter to a close, then, I want to ponder what Mike Davis (1993: 55) calls 'the forgotten architecture and casual detritus of the first nuclear war'. In his essay 'Dead West: ecocide in Marlboro country', Davis discusses the legacy landscapes of the US's nuclear proving grounds, and the fate of the '**downwinders**' – the victims of fallout from the testing of nuclear weapons. Like the 'casual detritus' of the space age, these littered landscapes invoke an *unheimlich* and heterotopic response. And like abandoned factories, derelict houses and depopulated ghost towns, there is something fascinating but also something dreadful about these nuclear ruins (on other contemporary ruins, see Edensor 2005). They at once signify the wastefulness of nuclear and space science, the passing of an era marked equally by promise and paranoia, and the careless bravado that infected the way the military-scientific-industrial complex viewed and treated the land (and its people). The architecture and landscapes of the Moon and the Bomb age stand as relics embodying both optimism and paranoia – features which cast a large question mark over what to do with them, how to feel about them, today.

Tom Vanderbilt's archaeological travelogue of atomic America, *Survival City* (2002), offers an insightful meditation on the meanings of these 'ancient ruin[s] of an accelerated culture' (25). US nuclear sites are depicted by Vanderbilt as 'place[s] where the future itself was tested' (43), and touring them today provides 'a portent of how the future was going to be' (47). In these relics, Vanderbilt writes, we see 'the past of a future averted' (100). Visiting these places, some now transformed into museums or heritage sites, others repurposed or just abandoned, is – as Vanderbilt concedes – a very peculiar form of tourist experience:

All wars end in tourism. Battlefields are rendered as scenic vistas, war heroes are frozen into gray memorials in urban parks, tanks and other weapons bask outside American Legion posts on suburban strips. That the Cold War, the so-called 'imaginary war' . . . was never actually fought . . . makes its tourism somewhat odd. This tourism curiously combines the 'what if' with the 'what was'.

(Vanderbilt 2002: 135)

What motivates people to want to visit these Cold War ruins? And what motivates the preservationists who want to museumify the Cold War? These questions were brought home to me very keenly during a visit to my neighbourhood Cold War visitor attraction, Hack Green Secret Nuclear Bunker in Cheshire (see http://www.hackgreen. co.uk). Entering the bunker, wandering the exhibits, is indeed a curious experience, full of imaginary memories and dense intertextuality, pulling up images from sci-fi and war movies. Room after room filled with obsolete technologies and paranoid infrastructure, bulked out by related paraphernalia and ephemera from the Cold War. The mood of the visit is sombre, at times even scary, though this is off-set by the screaming kitschness of other parts of the site – the canteen staff in military uniform, the weird mix of souvenirs for sale (including atomic bomb snow globes, reprints of the government's *Protect and Survive* manual, but also genuine cast-off military gadgets and oddments), the repeatedly tannoyed attack warnings. Even the exhibits themselves have a kitschness about them, not least in terms of the now-antiquated equipment on display. As Vanderbilt notes on visiting a former Cold War facility, 'the technology, advanced for its day, [now] seems whimsically outdated' (164). I was mesmerized by Hack Green, but also quite troubled. My boyhood enthusiasm for militaria clashed with my adult anti-war sentiments. And I couldn't get a reading on the other visitors, either – the families day-outing, the middle-aged couples grazing the exhibits. What cultural work is Hack Green doing?

Similar questions and tensions are raised in essays by Arthur Molella (2003) and Hugh Gusterson (2004a). Molella interrogates the problem of 'atomic museums' in the US, in terms of 'official' exhibits of American nuclear culture, usually located at former Cold War installations. These are authorized examples of museum, most funded in the past by the Atomic Energy Commission or the later Department of Energy, though contracted out operationally to private managers. The key problemic with these museums for Molella is partly a symptom of this funding legacy, and partly a symptom of their sitedness in the midst of communities whose pasts are intimately connected to Cold War culture. This inevitably compromises the stories their exhibits tell:

> Their views of the Bomb and the Atomic Age remain oddly distorted and veiled, revealing much about the imperatives and technical aspects of atomic bomb development but virtually nothing about their actual uses and unimagined destructiveness . . . Atomic museums have been captives to time and place, to their origins and to local cultures steeped in the crisis of war and, subsequently, potent Cold War ideologies.
>
> (Molella 2003: 211)[6]

Molella's main focus is a museum at one of the Manhattan Project's 'secret cities', Oak Ridge, Tennessee, somewhat euphemistically named the American Museum of Science and Energy (AMSE), where he encounters a powerfully coded exhibit of Bomb culture: 'Predictably patriotic in its presentations of the Bomb, AMSE displays its nukes proudly' (215). Nevertheless, Molella also experiences an 'uncertain mood' in Oak Ridge, partly as a result of the downsizing of the nuclear industries and the growing concerns with their health and environmental legacies – concerns at odds with the patriotic preservationist imperative. Moreover, in the post-9/11 climate, sites like Oak Ridge have been subject to heightened security and tainted by a distaste with displaying weapons of mass destruction, especially when the exhibits have been so one-sided, so overwritten with the Cold War logic of information management. While Molella concludes in a downbeat tone, that the fate of these sites signals the nation's inability to deal with the Bomb, he doesn't consider the ways in which museum visitors might interpret even heavy-handedly overcoded exhibits in resistant or oppositional ways. While the curatorial intent is to tell one kind of story, therefore, the 'text' of the exhibit is never totally closed, especially as visitors bring their own experiential and intertextual work to the museum.

Conflict over remembering the Bomb is also at the heart of Hugh Gusterson's paper on visiting 'ground zero for nuclear tourism' (2004a: 24) – Los Alamos, New Mexico, another key site for the Manhattan Project. Gusterson's trip, however, finds more contestation over the 'official' version of US atomic history told by the exhibits:

> Seeking to present the development of nuclear weapons as natural acts of history rather than a continuing spur to controversy, [nuclear museums] frame the history of nuclear weaponry in a way that seems designed to encourage awe for the achievements of the scientists who built the bombs while marginalizing questions about the political and environmental costs of the Manhattan Project and the ensuing arms race. But the feelings aroused by nuclear technology, even yesterday's nuclear technology, are not always easy to channel and contain, and the artefacts of the first nuclear age serve as magnets for partisans of opposed understandings of that era.
>
> (Gusterson 2004a: 24)

So, at Los Alamos Gusterson experiences conflict over the memorializing of the Manhattan Project, with anti-nuclear tourists equally drawn to these sites (even though any form of protest is forcefully policed). Like many contested forms of heritage, nuclear museums are in and of themselves complex and ambivalent: they have an authorized version of a story to tell, yet they are overshadowed by other stories, which constantly threaten to surface.

Of course, these sites are only one part of the legacy landscapes of the Cold War. Vanderbilt (2002) also focuses on unofficial, unclaimed, unpreserved relics – as well as those that continue to be shrouded in secrecy and denial. In the UK and USA, preservationist bodies are at work attempting to save nuclear heritage sites from destruction and decay. In America, the Atomic Heritage Foundation campaigns for significant sites across the country to be conserved and to become part of the heritage tourism itinerary; English Heritage recently published an extensive audit, *Cold War Monuments*, and is rolling out plans to protect Britain's nuclear landscapes from dereliction, with some of the most historically significant sites being designated as Scheduled Ancient Monuments (Cocroft and Thomas 2003). Other 'unofficial' groups of enthusiasts, such as Subterranea Britannica, are also engaged in audits of civil defence sites (see http://www.subbrit.org.uk). There is even in the US a realty market for Cold War buildings, with former missile silos and other facilities being sold as storage facilities, for industrial uses, and even as homes (see http://www.missilebases.com).

If the desire to preserve and museumify the Cold War represents an ambivalent form of collective working-through, where authorized versions of our recent history are presented but also contested (Woodward 2004), there are other related forms of heritage tourism that need to be set alongside the silos and bunkers, in a return to my twinning endeavours that underscore this chapter: space science tourism. If we are to read the space race as a powerful arm of the Cold War, then we need to think about the cultural work that these sites do. Marina Benjamin (2003) vividly describes a trip out to Cape Canaveral, Florida, which we might call 'ground zero' for US space tourism.[7] This trip is equally ambivalent, melancholic even, as Benjamin sees a landscape of downsizing and unfulfilled expectation. The dream of the space age was short-lived, and never delivered the economic prosperity that the area hoped for. Like the struggling nuclear towns such as Oak Ridge, the town of Brevard was the seat of a very short-lived gold rush, and equally quickly suffered the slump that so often follows boom: 'as I headed towards the Cape on a near-empty freeway, surrounded by scrub, what surprised me most was how so much history could have been so quickly erased. It was as if the Space Age had never happened' (Benjamin 2003: 10–11). Benjamin describes her trip to the Kennedy Space Center with immense disappointment, seeing it as too backward-focused, too Apollo-obsessed, though she later reflects that her motive for visiting the Cape was in itself partly nostalgic:

> What I had been attempting to recapture at the Cape was something more than the sum of the Space Age's parts – something more than a heritage-centre-style montage of rockets and gantries, architecture and landscape. I was after inspiration. I wanted to recapture the sense of hope with which all things related to space filled me as a child, as it filled the adults around me. I wanted to remember the taste of dreaming about our Space Age future.
>
> (Benjamin 2003: 17–18)

This sense of disappointment, of utopia and progress replaced with nostalgia and heritage, seems to contradict Constance Penley's (1997) contention about space and the American imagination; like the Cold War museums, these space heritage sites offer up 'a portent of how the future was going to be', like Vanderbilt says, but equally profoundly a reminder that the future of the past has not turned into the present. As powerful elements of the broader sense of technostalgia that also includes 'classic' computer collecting, retro space-age pop music, the pulp futurology of sci-fi and the whole ragbag of retrofuturist kitsch (see Brosterman 2000; Holliday and Potts 2006; Taylor 2001), these legacy landscapes of the Moon and the Bomb are therefore acutely evocative of both the promises and the limits of science, technology and culture, and of the gap between the present as it is and the present imagined as the future in the past.

Further reading

Benjamin, M. (2003) *Rocket Dreams*. London: Chatto & Windus.

Cocroft, W. and Thomas, R. (2003) *Cold War: Building for Nuclear Confrontation 1946–1989*. Swindon: English Heritage.

Gusterson, H. (2004) *People of the Bomb: Portraits of America's Nuclear Complex*. Minneapolis MN: University of Minnesota Press.

Henrikson, M. (1997) *Dr Strangelove's America: Society and Culture in the Atomic Age*. Berkeley CA: University of California Press.

Light, M. (1999) *Full Moon*. London: Jonathan Cape.

Light, M. (2003) *100 Suns*. London: Jonathan Cape.

McCurdy, H. (1997) *Space and the American Imagination*. Washington: Smithsonian Institution Press.

Newkey-Burden, C. (2003) *Nuclear Paranoia*. Harpenden: Pocket Essentials.

Nye, D. (1994) *American Technological Sublime*. Cambridge: MIT Press.

Penley, C. (1997) *NASA/TREK: Popular Science and Sex in America*. London: Verso.

Spufford, F. (2003) *Backroom Boys: The Secret Return of the British Boffin*. London: Faber & Faber.

Vanderbilt, T. (2002) *Survival City: Adventures among the Ruins of Atomic America*. Princeton NJ: Princeton Architectural Press.

Notes

1 Penley reads this claim sceptically, writing that maybe 'the students wanted or needed that irony [of watching the disaster in science class] to rationalize a changed relation to science and technology, as well as to their own futures' (44).

2 Although NASA may be currently refocused on Mars, the dream of space tourism is being kept alive by the US government, but shifted into private, entrepreneurial hands: on 4 March 2004, the US House of Representatives approved legislation designed to promote the development of the emerging commercial human space flight industry. H.R. 3752, The Commercial Space Launch Amendments Act of 2004, if approved by Senate, will put in place a regulatory regime to promote the passenger space travel industry. See http://www.spacefutures.com/home.shtml; on future potential for lunar tourism, see Collins (2003).

3 Ironically, in terms of popularity, *Dr Who* began to lose out in the early 1980s when the show was scheduled against a much more American time-traveller, *Buck Rogers in the 25th Century* (Cull 2001). The 'regeneration' of The Doctor in 2005 met with some press scepticism, given the high expectations of today's sci-fi fans used to big budget blockbusters, but the series proved a critical and popular success, managing to maintain something of the 'feel' of previous series.

4 It is worth noting that Pillinger also played a key role in the Martian microbe story, by being responsible for verifying that AHL84001 was in fact of Martian origin, though

he remained agnostic over the question of whether the 'fossils' in the rock were evidence of past life on the planet (Appleyard 2003).

5 On 1 April 2004, as an April Fools' jape, the European Space Agency auditioned for a space dog to go to the International Space Station. Excited dog owners brought in their pets, who were made to stand in a 'launch simulator' to test their aptitude for space flight. For details (in Dutch), see http://www.spaceexpo.nl/laika_content.html. Thanks to Wendy Murray for emailing me about this.

6 Molella is no stranger to this problematic, having curated a controversial exhibit about the Manhattan Project at the Smithsonian Institute, *The Crossroads: The End of World War II, The Atomic Bomb and the Origins of the Cold War* (see Gieryn 1998).

7 We might argue, in terms of symbolics, that in fact the landing site on the Moon is 'ground zero' for the space age. There have been calls to turn these sites into archaeological monuments (see Fewer 2002).

6 | UFOLOGISTS, HOBBYISTS AND OTHER BOUNDARY WORKERS

One of my favourite cartoons shows a fortune-teller scrutinizing the mark's palm and gravely concluding, 'You are very gullible.'

(Carl Sagan)

Walk into any high-street bookshop and you are likely to find two categories of books that are being granted considerable shelf-space. In the shops I mostly frequent, one of these is labelled Popular Science, and the other Mind, Body & Spirit. On these packed shelves you can find all kinds of books – both categories are still caught in a publishing boom which has been going on for some years now – on topics ranging from self-help to self-sufficiency, and from the story of the number zero to the physics of *Star Trek*. It is this last type that I want to spend a moment or two thinking about here, as a route into discussing the implications of Ziman's (1992: 15) simple-yet-complex statement that 'what counts as science is defined very differently by different people – or even by the same people under different circumstances'.

As we have already seen, the idea of 'popular science' is a broad and contro-versial one today. There are debates about how to make science more popular, but also about how to save it from populism. There are also, as we saw in Chapter 4, debates about the ways popular culture depicts science – and one response to this has been the publishing of *Science Of . . .* books. Depending on popular cultural texts most prominent at any time, these books don't lag far behind. *The Physics of Star Trek* (Krauss 1995) is a very well-known example, joined as trends in popular culture moved on by *The Science of the X-Files* (Cavelos 1998) and *The Real Science Behind the X-Files* (Simon 2001), *The Science of Star Wars* (Cavelos 1999), *The Forensic Science of CSI* (Ramsland 2001), *The Science of Harry Potter* (Highfield 2002) and *The Science of the X-Men* (Yaco and Haber 2004). These have also been joined by broader-ranging titles, such as *The Science of Aliens* (Pickover 1999) and *The Science of Superheroes* (Gresh and Weinberg 2002). On closer inspection these books fall into two distinct camps. The first borrows the attention on a popular cultural text, turning it to the service of science pedagogy (as well as, arguably, cashing in); the other is concerned with disproving or correcting the *misrepresentations* of science in popular culture, with showing that some of the things in shows and films like these are scientifically implausible, even impossible. This distinction is one important kind of boundary work, and boundary work in technoscience is this chapter's focus.

The trouble with boundaries

The issue of what science, technology or technoscience is – that is, how to define it – brings up immediately the issue of boundaries. Boundaries are at the heart of how we understand things in the mindset we might call modernity. We have a way of naming and classifying things that depends on a twofold division and mutual exclusivity: it is this, or not-this. We don't like things that are kind-of-this, not-quite-this or possibly-this. The clearest form of classification, then, is binary and oppositional, with no room for ambiguity, no maybes. Bruno Latour (1993) has written about this in some detail, in *We Have Never Been Modern*. Modernity, he says, is all about simplification, classification, ordering – it's what we might call taxonomic tidying, or what he calls **purification**. But Latour detects a paradox in modernity, a paradox that we can usefully use to talk about science and technology. While on the one hand we have the practice of purification, which creates pairings of 'entirely different ontological zones' (Latour 1993: 10) – such as human/non-human, nature/culture, male/female – modernity has also been about the production of hybrids which straddle, blur and mess with this tidying (this process he calls **translation**). So we want to keep things simple but keep making things more complicated, and the conflict between the two imperatives is most keenly felt, most clearly seen, in the border zones. At the border, then, we can witness the clash between purification and translation, and begin to explore how, as Laura Nader (1996: 3) writes, 'science is not only a means of categorizing the world, but [also] of *categorizing science itself* in relation to other knowledge systems that are excluded' (my italics).

This idea has been discussed by many critics interested in the sociology of science. Thomas Gieryn (1995, 1999), for example, has provided one of the best-known elaborations of 'boundary work' in terms of scientific knowledge. His query concerns the authority or legitimacy of scientific knowledge – how some things get to be called truths, who decides this, and how the status of truth is managed. Of course, those of you either old enough to have caught it first time, or with access to the cable and satellite channels that are now its home, will recognize the issue of the boundaries of truth as having been played out through the popular TV series *The X-Files*, whose strapline was 'The truth is out there'. *The X-Files* messed around with ideas of truth, not only in terms of science but also in terms of regimes of political truth, through its focus on conspiracies, cover-ups and so on (another of its slogans being 'Trust no one').

For a time, the popularity of the *X-Files* got some scientists mad, as they felt it was misrepresenting scientific knowledge, scientific practice and scientific truth (see Allan 2002 for an overview of this moment). *The X-Files* was blamed for assorted outbreaks of irrationality – for upsurges in belief in UFOs and alien abduction, for undermining the authority of mainstream science by privileging fringe, pseudo- and anti-scientific ideas (one more slogan from the show: 'Deceive, inveigle, obfuscate'). Given the short attention-span of popular culture, noted in the introduction to this book, revisiting the *X-Files* debates might leave a lot of readers clueless, as the show and its impacts recede into memory. It should be remembered, though, as a significant recent moment in boundary work around popular science, and one that continues to leave traces at that particular boundary.

Boundary work is, in Gieryn's formulation, an on-going task; moreover, it is not immutable, fixed in stone – as Ziman says in the quote earlier in this chapter, different people draw the line in different places, and the same person might even do this at different times, or around different issues. The porosity and elasticity of the boundary between science and not-science does have limits, of course, but those limits are culturally (and historically and geographically) contingent; they are local and situated. As we shall see, in another moment of untidy tidying, we can divide the not-science category into some subsets, that we can name popular science, pseudoscience, fringe science and antiscience.[1] As we shall also see, continuing to interpret these knowledges as types of not-science, and thus as types of non-knowledge (Featherstone 2002), contributes accidentally to the boundary work of keeping science separate and special. It's something I shouldn't really be doing in a book like this. My defence is that to understand boundary work, and the epistemic spaces that are bounded in or bounded out, we have to make visible those very boundaries we want to argue against or about. My use of terms like 'fringe' or 'pseudo' refers to the boundary-policing that labels certain thoughts and activities in those ways, not to their absolute positioning on a grid with 'real science' at its centre – though that is precisely the positioning that those doing the border patrols want to preserve or produce. Trying to work a version of symmetry here, I want to try to resist reproducing the boundary work I am discussing, though I am talking about these different categorizations of 'not-science'.

Perhaps we should begin by thinking about the term 'popular science', and what meanings this may have in the context of boundary work. As already hinted at a number of times in this book, this term, which names and clusters together heterogeneous titles on the shelves of bookshops, has both positive and negative spins. The positive spin reads 'popular' as popularization, as making things accessible, of engaging the public's interest. The book that emblematizes this, and which has a key (if controversial) role to play in the stimulation of the publishing explosion in pop science books, is of course Stephen Hawking's (1988) *A Brief History of Time* – the book that, so the folk story goes, everyone bought but no one read, the book that made its author a media star and that re-enchanted science for the reading public.[2] A more recent offspring, Bill Bryson's (2004) *A Brief History of Nearly Everything*, does even more, and is written by a non-scientist (those boundaries again!), perhaps thereby solving the 'unreadness' of Hawking.

The negative spin slides popular into populist, meaning today dumbed down as much as anything else (see McGuigan 1992). Andrew Ross (1991) notes the homology between judgements about science and taste cultures, seeing the most troublesome category in both cases lying in the middle-ground – in science/not-science terms he sees this as New Age science, and in taste cultures it is the middle-brow. For Ross, popular science is the low-brow; and for other commentators, too, populism means appealing to the masses, and resembles what Carl Sagan (1997) dismissed as the 'tabloidization of science' in his polemic defence of scientific rationality, *The Demon-haunted World*. Popular science does indeed sometimes incorporate tabloid tales, or attract tabloid attention, as we shall see. But there is a broader ambivalence at work here, which is to do with the 'specialness' of science – itself a kind of taste discourse. If science becomes popular culture, available to 'non-scientists', will that turn everyone into a scientist, and so degrade the status of scientific knowledge?

This difference is captured in the comparison between Hawking and Bryson as science popularizers. One is deeply embedded in scientific knowledge and practice, the other a self-confessed novice, using little more than a curious mind and some journalistic skills. Both are trying to explain stuff that's hard to understand, and both recognize the value of such explanation, but also some of the compromises. Even if *A Brief History of Time* was regarded as incomprehensible to most people who bought it, becoming instead a kind of status symbol and folkloric object, to condense down all the science into a slim book was an enormous feat of simplification, and relied on writing techniques not commonly used in scientific publications, such as metaphor (even though, as we saw in Chapter 2, scientists make use of analogy in their own thought processes, for example to comprehend geological time). The science of boiling an egg, discussed in Chapter 1, is similar (but also strikingly different).

Bryson's populism belongs to a different tradition, closer to a pop-culture approach to science; he defers to 'experts' for the science bits, and then translates these for his readers, who he assumes know only about as much science as he does. So his writing is part of a parallel tradition of popularization, produced by authors and journalists 'outside' of science (Francis Spufford, author of *Backroom Boys* (2003) discussed in Chapter 5, is another example among their many and growing numbers).

One of the biggest areas of debate about the approach to and effects of popular science or science popularization has centred around the 'displaying' of science in museums, science centres and other similar visitor attractions (see Allan 2002; Fahey *et al.* 2005). In the context of the so-called 'experience economy', when people are searching out ways to pass their leisure time interestingly and productively, science museums and centres have proliferated – and this proliferation can also be seen as part of the science popularization agenda. 'Hands-on' and interactive exhibits attempt to conjure some of the wonder of science, with participation also helping to demystify or 'ordinari-ize' science.

Johanna Fahey *et al.* (2005) discuss this approach to popularization, and its limitations, in Australia. Their essay, 'A Taste for Science' focuses on the problem of young people's lack of engagement with and participation in science. This is framed as something of a national emergency in Australia (and indeed, elsewhere), given the projected requirements of the knowledge economy. Without science-literate workers, the chances of succeeding in knowledge-based sectors look slim. One prominent way of recruiting young people back to science – given the apparent inability of formal education to make science 'sexy' – has been through an emphasis on interactivity, most commonly through science-themed visitor attractions known as interactive science and technology centres (ISTCs). But, as Fahey *et al.* demonstrate, ISTCs are based on an outmoded model of science display, but more importantly on an outmoded model of young people: the stress in ISTCs on 'family fun' is exactly the kind of idea guaranteed to send a shiver down any young adolescent's spine: family fun is alright for *kids* (and parents), but it's too childish for 'youth'.

But the pressing issue of cultivating a taste for science among young people has also led to the co-option of other symbolic resources, including those associated with ordinari-ization and lifestylization (see Bell and Hollows 2005). In Australia, popular science has found the lifestyle media a youth-friendly platform on which to stage science's re-enchantment. Fahey *et al.* explore this through a discussion of *Sleek Geeks*, a science radio show aimed at young people in Australia that knowingly addresses its implied audience, using humour and 'yoof-speak', to produce 'ordinary science'. Playing on the image of the scientist as geek – a recurring symbol of the unsexiness of science in popular culture (see Chapter 4) – *Sleek Geeks* has blended youth lifestyles and science in a way that is miles away from the 'hands-on but out-of-touch' family fun of the ISTCs. One key strategy used by the show's presenters, much like that used in *A Short History of Nearly Everything*, is the de-emphasizing of expertise. As Frances Bonner (2003: 48) discusses in the context of BBC archaeology show *Time Team*, having an 'eternally naïve' presenter gives science pedagogy a way to render science digestible, that is, *ordinary* – a trick to make the moments of expository science seem less contrived than they can be in shows like *The X-Files* or *CSI*.

Fictional representations of science also have a powerful but ambivalent popularization effect, as we saw in Chapter 4. While shows like *The X-Files* attracted a lot of negative attention from scientists, for example, one aspect of the show that was seen positively, at least by some people, was its depiction of a female scientist in the central character Dana Scully (Wilcox and Williams 1996). In playfully upturning the association of males and scientific rationality, Scully was the show's sceptic and scientist, a foil to Fox Mulder's 'irrational' belief in alien abduction and government conspiracy. An extratextual effect of this narrative commented on at the time was that it reopened the door to science for girls and women who had previously been excluded or had excluded themselves from education or employment in 'male' science.

So TV can make science 'sexy'. More recently in the UK, an upsurge in university enrolments on courses in forensic science is anecdotally attributed to the success of the *CSI* franchise. This set of three American shows – *CSI: Crime Scene Investigation*, set in Las Vegas, plus more recent spin-offs *CSI: Miami* and *CSI: NY* – centre-stage the role of science in police work, with much of the action occurring in labs. (A similar effect a few years back reputedly connected the hit show *Cracker* with renewed interest in criminology and psychology.) Now, while there is inevitably a folkloric dimension to these tales, they should nevertheless make us mindful of the potential of popular culture to popularize science. At the same time, however, we might reiterate Gregory and Miller's (1998) comment, seen in Chapter 4, that popularizing science doesn't always make it more popular – hence some of the ambivalence expressed by scientists about aspects of the project of popularization. One prominent strain of this ambivalence points to the ways that popular science blurs the boundaries between 'proper' scientific knowledge and practice and its disreputable relatives, fringe and pseudoscience. Popular science is castigated for propagating a public *misunderstanding* of science, through oversimplification, through sensationalism, or through wrongly mixing science with the occult, paranormal, alternative (Sagan 1997 remains one of the best-known critiques of popular science; for critique of Sagan, see Allan 2002; Penley 1997). It is to these borderlands that I now want to turn.

The outer limits

Earlier in this chapter I took you to the bookshop, noting the heaving shelves of the Popular Science and Mind, Body & Spirit sections. Booksellers are here performing their own boundary work, symptomatic of the broader business I am discussing here; for while you will find Hawking or Bryson in the Popular Science section, along with books by writers such as Stephen Jay Gould and indeed Carl Sagan, and sometimes history and even sociology of science, books on astrology, ufology, crystal healing, etc., are shelved under Mind, Body & Spirit, increasingly a euphemism for 'New Age' beliefs and practices (see Hess 1993). Some books and topics are 'sciencey' enough, clearly, while others aren't. So this everyday demarcation, in terms of what David Hess (1993) calls the 'scientificity' of books and their shelving (and also thereby their purchasers, who perform their own boundary work by gravitating to some shelves but not others), contributes to the broader boundary work I am talking about here.

One boundary based on scientificity, used to categorize non-mainstream scientific knowledges and practices, makes a split between pseudoscience and fringe science.[3] I have spent a lot of time pondering this, and although I feel it is an untidy, even at times untenable split to make, I want to discuss it here as an example of the boundary policing that takes place even within and among things that mainstream science refuses to countenance. Both fringe and pseudo-science can be regarded as forms of 'alternative science', and are distinguishable from antiscience, at least for me, in terms of their continued insistence on scientific knowledge and method, no matter how vulgarized these are seen to have become by mainstream science. Here is the closest I can come to some sort of definition and explication. Pseudoscience refers to the application of what are claimed as forms of scientific knowledge and method to the exploration (and proving) of things that mainstream scientists don't regard as science. Here I would include lots of things that end in -ology, the suffix that signifies scientificity, such as astrology, parapsychology, ufology, scientology (although this mixes pseudoscience and religion; see Davis 1998), and also things like Fortean science. I take fringe science to mean knowledges and practices produced and performed by scientists that contest or stand outside of mainstream science. Andrew Ross (1991: 19) defines it as science done by people 'within the legitimate scientific community whose work is locally contestatory and thus marginalized or suppressed'. Fringe science can sometimes be slotted into the Kuhnian frame of paradigm shifts, as in one example I mention below, that of 'warm-blooded dinosaurs'. Other fringe sciences remain outside, never contributing to a paradigm shift. Examples I will draw on here include panspermia and alternative medicine. In fact, there's a further splitting here, between aspects of fringe science that attempt to enter the mainstream by contesting it (such as the already mentioned panspermia theory), and those which maintain a willful outsiderhood – here I am thinking of some parts of the **hacking** subculture, for example, or so-called gonzo science.[4]

Now, before you start shouting, let me reiterate how uncomfortable this split feels. As Ross (1991) has shown so clearly, for example, what we think of as New Age science flits between the two. And in terms of their connection to my earlier discussion of popular science, it should be noted that some areas of fringe and pseudoscience maintain their outsiderhood, their unpopularity, here too – tucked into the esoteric and arcane 'marginals milieu';[5] others meanwhile transgress into the popular, for example through publication as popular science – as in the case of Hoyle and Wickramasinghe's (1983) panspermia book, *Evolution From Space*, discussed below. In what follows I am going to abandon my provisional attempt to tidy these activities and thoughts into fringe science and pseudoscience, because the boundary here is just too fuzzy: what, for example, connects or separates crystal healing and the **Atkins diet**, or Atkins and ufology, or ufology and crystal healing? While they might be seen as elements in a broader mosaic of marginalized knowledges and practices, even for those like Latour (1993: 122) (and me) who have 'a perverse taste for the margins' there are both similarities and differences here. But the point here is not to judge, to perform our own boundary policing; using the symmetry of the Strong Programme discussed in Chapter 2, our job is to explore the *cultural work* that these knowledges and practices perform, and also to explore what's at stake in the boundary work that preserves a centre–margin split in science and technology.

For many commentators, such a split is not only unreasonable, it is counter-productive. These writers argue that much of what we today see at the centre of science and technology in fact emerged from the margins, reflecting the punctuated equilibrium of Kuhnianism and the heroized figure of the 'maverick' scientist (see Chapter 4):

> Really significant advances have always grown out of the revelations of independent thinkers and tinkerers who were not learned enough to know that they were violating the laws of physics or any other branch of science.
>
> (Eisen 1999: 2)

Eisen's (1999) *Suppressed Inventions and Other Discoveries* is a reasonably representative example of one kind of fringe science text (for commentary, see Bell 1999b). It provides a conspiratorial compendium of the field, and describes some things that to 'conventional' scientific wisdom seem unthinkable – cars with petrol engines converted to run on water, treatment of cancer using 'somatids', anti-gravity devices, crushed rocks as fertilizer. Given its embedded-ness in conspiracy discourse, many of the tales it tells end in mysterious disappearances, strange silences – but that's OK, because in the conspiracist's world-view, lack of proof is in itself proof; proof of cover-up (Fenster 1999). It is adamantly of the fringe, disavowing the 'scientific establishment' for narrow-mindedness, and for being puppets of the state or multinationals – thus repeated successes in developing alternative sources of fuel are silenced by petrochemical interests pulling the strings of scientists who use the veil of scientific objectivity to dismiss anything unorthodox. Failing that, maverick scientists can be discredited, spooked, even disappeared.

However, not all instances of fringe science follow the same logic. The once-eminent, knighted British astronomer Sir Fred Hoyle provides us with an interesting case study in fringing, thanks to his attempts to promote the theory that, as Jane Gregory summarizes,

> interstellar dust grains are coated with organic material that is swept up and processed in comets; on cosmic timescales, an orbiting comet experiences the range of conditions needed to turn simple carbon-based compounds into complex biomolecules. Life, in the form of these bio-molecules, first fell to Earth from a passing comet. Biological material – including bacteria and viruses – is still arriving on Earth in this way, which accounts not only for the epidemiology of diseases such as flu, but also for the genetic variation that leads to evolution.
>
> (Gregory 2003: 25)

In her telling of the tale of Hoyle's 'excommunication' from mainstream science, Gregory points up a number of important things. She highlights the role played by a range of media – scientific and popular – in promoting and contesting Hoyle's ideas: Hoyle himself wrote science fiction alongside both 'academic' and 'lay' science books, newspaper articles, and so on. Critics also used mixed media, venting their scientific spleen via book reviews, outside the controls of other forms of scientific publishing (peer reviewing, etc.). Vitriolic and often very personal reviews furthermore 'did boundary work in so far as they served to separate Hoyle from the collegiality of the scientific community' (Gregory 2003: 36) – they constructed a common 'we' in opposition to a lone 'him'. Professional ostracization thus serves to reaffirm the orthodoxy of science, by consigning others to the status of unorthodoxy (more on this later). However, as Hoyle found his ideas increasingly marginalized by science, he made tactical use of the pop science (we might even call it 'pulp science') publishing boom, using the popular science press to publicize ideas considered unorthodox by other scientists.[6]

Part of Hoyle's (and his frequent co-author, Chandra Wickramasinghe's) argument stemmed from incredulity towards Darwinism. This incredulity led them into further hot water when in 1986 they published a book alleging that a fossil of the so-called 'missing link' between dinosaurs and birds, *Archaeopteryx*, was a clever fake. While fossil-faking has a long history, claiming that so famous a fossil – and so famous a proof of Darwinism in practice – was a hoax, cast Hoyle even further to the fringes, his ideas conclusively discredited.

Archaeopteryx occupies a central position in newer thinking about the relationship between dinosaurs and birds (Mitchell 1998), which also connects with a miraculous transformation in our understanding of dinosaurs that has impacted on their reconstructed morphology, tales of their evolution and of their descent. This is the shift from seeing dinosaurs as totally analogous to reptiles, including their being ectothermic (cold-blooded), towards seeing them as endothermic, or warm-blooded. In a lively account of this transformation, Adrian Desmond (1975) describes how dinosaurs were misrecognized as reptilian based on fragmentary fossil finds; and, like modern-day reptiles, they were assumed to be cold-blooded, needing external warmth from the sun. As Desmond shows, the modern tendency towards tidying via taxonomy has here proved troublesome, as new discoveries and theories questioned the grouping of such diversity into the class *Reptilia*:

> The classification has failed to keep pace with our changing ideas, resulting in an anomalous situation whereby the reptilian class contains creatures as diverse as furry, intelligent, endothermic flyers; archaic, sprawling, small-brained ectotherms (lizards and turtles); giant, erect, terrestrial endotherms; and swift-running, extremely sophisticated dinosaurs, at least one of them (*Archaeopteryx*) with feathers.
>
> (Desmond 1975: 198)

As W.J.T. Mitchell (1998) adds, in his epic cultural reading of the dinosaur, this diversity has made a nonsense of the very term 'dinosaur' – which was always equally inappropriate and inaccurate as the name allegedly first given to the first fossil dino bone, discovered and described by famed natural historian Dr Robert Plot in the seventeenth century, and named in the late eighteenth century for its resemblance to a part of a man's anatomy, *Scrotum humanum*. (Plot identified it as a thigh bone, but not as fossilized, and the other fossils he unearthed he attributed to natural, chemical processes, with ammonites for example being the result of the reaction of two salts; see Edwards 1976.)

Let's stay with the human body for a moment longer, but turn from the scrotum to the stomach, from one doctor to another, and to a contemporary controversy over the effects of what we eat on our health and girth. Here's its leading advocate explaining his particular challenge to scientific orthodoxy: 'since the late 1970s a simplistic theory of weight loss that focuses obsessively on dietary fat and by implication teaches old, stale doctrine that gaining or losing weight is just a matter of calorie consumption has ruled the roost' (Atkins 2003: 1). Yes, it's Dr Robert Atkins, promoter of the eponymous diet that shuns carbohydrates rather than calories, that works by 'metabolic advantage' and by regulating insulin, and that has been incredibly popular and incredibly controversial here in the UK, as it has in the US and elsewhere.

Atkins is an enthusiastic proselytizer of his 'revolutionary' approach to dieting, and he knows that the business of convincing people in this field means appeals to scientific rationality, but also to the cultural capital of outsiderhood: 'there's sound scientific evidence for what I'm telling you now that is largely being ignored', he writes, adding that his ideas have 'gotten a bum rap'. He conscripts dieters into adding to the proof, asking them to have blood tests to illustrate the 'irrefutable, numerical improvements' to their weight and health (Atkins 2003: 3). His book goes on to provide an account of the biochemical processes that the Atkins diet produces in the body, which revolve around ketosis, also called lipolysis or 'the dissolving of fat'. In an interesting exemplar of boundary work here, a quick bit of Googling shows ketosis listed on the UK National Health Service (NHS) website as a potentially harmful symptom of metabolic disorder, but on countless weight-loss websites as a positive health-benefit of low-carb dieting. The anti-Atkins site www.attkinsexposed.org, meanwhile, provides pages debunking the 'faulty science' behind the claims. The Atkins diet has proved phenomenally successful, and popular. Lots of scientists say it is 'faulty', even harmful. Advice swings in one direction, then the other. Weight-loss seems to have hit a moment of postmodern relativism, full of radical uncertainty. What, and who, should we believe? And what should we be eating?

Of course, Atkins belongs to a much broader body of contestatory medical knowledge, which includes myriad versions of so-called alternative and unorthodox medicine (Saks 2003). There has been, in many countries in the West, a heated and protracted debate between mainstream and **complementary and alternative medicine** (CAM), which is still far from being resolved. The turn towards such alternatives is read as a symptom of the failure of the 'medical establishment' and its narrow range of methods, its drug dependency, and its failure to ameliorate so many people's health problems. Doubt and critique hang over mainstream medicine, as we saw in the UK with the issue of the safety of the **MMR** (mump, measles and rubella) triple vaccine. As Mike Saks (2003) writes in his tidy history of this contest, a significant boost to such fringe practices was catalysed by the 1960s counterculture, which often mobilized a critique of science and technology in their current forms, and searched out alternatives from outside of contemporary Western culture. This counterculture gave shape to many species of fringe knowledge and practice, including New Age-ism and, of course, modern personal computing (Ceruzzi 1998;

Davis 1998). Lack of faith and trust, dissatisfaction with the narrowness of the mainstream, and the greater availability of alternatives, have together radically transformed the landscape of medicine, even for many people who would never associate themselves with hippies or the New Age.

The issue of *availability* of alternatives is, of course, of paramount importance. One of the cornerstones of the counterculture was the publicizing of different, fringe activities and beliefs, through publications like the *Whole Earth Catalog* (Binkley 2003). In fact, as Irwin Weintraub (2000) shows, alternative presses and now web-based resources are very significant channels for the dissemination and consumption of non-mainstream medical and other fringe science information, even if it allows a bewildering clash of divergent viewpoints.

While some aspects of alternative medicine are currently being 'tested' – by mainstream medical science, on its terms – and are therefore moving to a less marginal position, other zones of fringe or pseudoscience seem destined to remain on the outer limits, never to gain any veneer of scientific credibility (though paradigm shifts *do* occur). Iconic in this regard is astrology, described by Richard Dawkins as 'a debauching of science for profit' (quoted in Willis and Curry 2004: 95) and as 'a chronic and weeping sore' to many scientists (Gregory and Miller 1998: 55). Why such strong adverse reactions? First and foremost, because astrology is *popular*: horoscopes in particular are widely published, and read by millions. Second, because there is popular confusion, or slippage, between astrology and astronomy (Sagan was particularly irked by this). Third, it is constructed as plainly irrational, as fakery: how can the movement of planets determine individuals' life events, depending on their birth date? Fourth, who are these self-constructed 'experts' who can divine such information? Moreover, it's not just scientists who have criticized astrology and its believers. Cultural theorist Theodor Adorno (1994) lambasted astrology for passifying people, for making them see their fate as determined by extraterrestrial events – an analysis that chimes with his broader critique of popular culture's stupefying effects (for a discussion, see Bell and Bennion-Nixon 2001). Sagan (1997) rewrites this slightly more sympathetically, noting that astrology must fulfill social needs unmet from other sources – including science. Nevertheless, he is able to offer a paragraph of refutations of astrology that are unequivocal: astrology is straightforwardly a pseudoscience. Willis and Curry (2004) provide a defence of astrology, or a debunking of the debunking, through showing that 'scientific proof' is itself a conjuring trick (or a con trick). In a familiar critique – something that even Sagan concedes – they write that much of science couldn't pass the validity tests that it demands astrology conforms to.

Of course, this is part of the tight spot the so-called pseudosciences are put in. They can contest the validity of such validity tests, challenge the status of scientific authority – if they are happy to remain on the fringes. And, indeed, some branches of fringe and pseudoscience are happy with, even proud of, their outsiderhood, as we've seen. Some want to use that very outsiderhood to stage a critique of the mainstream and its values and beliefs, as in the case of the counterculture or alternative science. And yet others want to propose new sciences; to make a claim on the territory of science for knowledges and practices heretofore excluded. In this latter category belong, I think, the last two examples I want to talk about here: Fortean science and ufology.

Forteanism takes its name from Charles Fort who, in 1919 published *The Book of the Damned*:

> a loosely organized collection of observed events that, he argued, modern science had shunned simply because they were unexplainable in light of current theory, or were improper in light of turn of the century sensibilities . . . Reports of yellow rain, hailstones the size of elephants, falling fish, intuition, fakirs, stigmata, prostitution, rape, animal mutilations, devils and angels were all, Fort considered, worthy of resurrection from the dusty pages of journals and proceedings that he had poured over in the New York Public Library and, later, the British Museum Library.
>
> (Dixon 2005: 1)

As Deborah Dixon notes, Fort's ideas would come to influence subsequent generations equally interested in esoterica:

> His *Book of the Damned* has since been adopted as the foundational text of what has come to be known as Fortean science, composed of the critical examination of the 'paranormal' via the interrogation of eyewitnesses, archival research and fieldwork, and practiced primarily by those operating outside of the academic mainstream. Fortean science, then, can be located in what Fort himself might have termed a 'purgatory': though Forteanism questions the authoritative status afforded those scientific explanations that claim to reveal the systematic operation of laws, it simultaneously maintains a critical distance from those 'paranormal' revelations that seek to accomplish the same. Fortean science is accepting of the anomalous as a significant research focus, but is committed to the continual suspension of belief.
>
> (Dixon 2005: 1–2)

Contemporary Forteanism resurrected these interests, mainly via the magazine *Fortean Times*, which made its own miraculous transformation, blossoming on the back of the *X-Files* boom in interest in the 'paranormal': it turned from a small-run, low production values member of the marginals milieu to a glossy full-colour magazine sold over major news-stands. Taking its inspiration directly from Fort's life and work, *Fortean Times* reports unexplained events of all kinds, from UFO sightings to stigmatics, from divining to cryptozoology. As Dixon notes, the Fortean approach is not a wide-eyed acceptance of all these things at full face value; it depends on a 'critical distance' and thereby attunes itself to scientific practice, even though this distance is matched by an acceptance of the anomalous as a legitimate research subject.

This makes Forteanism a classic case of pseudoscience, in terms of how that label is used to demarcate the fake from the real. For those policing the mainstream's borders, Forteanism is even worse than plain belief, since belief can be translated into hoodwinking, as Adorno suggests about astrology. But by donning the mantle of investigative scientists, Forteanists are seen to be appropriating science, confusing the gullible public with their claims to truth and so on. By looking and acting like scientists, even if the things they study and the ideas they profess can be dismissed as bunkum, pseudoscientists are meddling with the status and image of science.

Jodi Dean (1997) writes nicely on this issue in the context of ufology, noting how it necessarily developed by emphasizing mainstream notions of truth and evidence – it had to do this, if it was ever going to convince sceptics of the *real truth*, based on the real evidence:

> Like science and law, [ufology] appealed to evidence. In order to defend the credibility of UFO witnesses, moreover, researchers appealed to precisely the sort of evidence they assumed would be acceptable to scientists and lawyers. Thus, they tended to reinforce official assumptions about who or what can be credible. Because ufology wanted to convince political and scientific authorities of the truth of its claims, it accepted their standards and criteria even as it resisted official efforts to monopolize evidence and discredit witnesses.
>
> (Dean 1997: 53)

In stringently policing its own inner workings as a way to try to gain legitimacy from the scientific mainstream, in the 1960s ufologists developed tests to measure the reliability of witnesses, which included physical and psychological assessments; as Dean writes, ufology 'redeployed truth itself' (56) in an effort to bolster the truthfulness of its truth claims. To strengthen its truth claims, ufology also deployed the props of mainstream science, as evidenced by the account of the well-known 1992 alien abduction conference at MIT (Bryan 1995) and by what Dean calls the 'ordinariness' of its publication *The Journal of UFO Studies*: 'like other academic journals, it employs a "double-blind, double-referee" system' (65), and has academics from 'respectable' universities on its board (though she notes that these are often marginalized in their institutions). While at times this aping of mainstream science seems to come with some postmodern irony mixed in – as in, for example, *The Field Guide to Extraterrestrials* (Huyghe 1996), a kind of ET-spotters' guide – even this underlines its veracity by being 'based on actual accounts and sightings'. Dean's charting of ufology's mainstreaming ends with an optimistic conclusion, that borders can be subverted through this kind of boundary work:

> Installed as practices for producing a knowledge that science disavows, heretofore reasonable procedures take an alien form. As the criteria for legitimacy are themselves abducted, the mainstream, the serious, the conventional, and the real become suspect. Not only does the UFO discourse cite scientific standards of objectivity, impartiality, critical debate, and consideration of alternative hypotheses, it also provides a location for the redeployment of those standards against institutionalized science.
>
> (Dean 1997: 65)

This leads to what she calls the 'fugivity' of truth – but also reinstalls the *promise* of truth: as *The X-Files* kept on saying, the truth *is* out there, somewhere (Bell and Bennion-Nixon 2001). Of course, the development of a scientific strand of ufology does not provide the only means by which people make sense of UFOs and aliens. Ufology was also engaged in internal border wars with alternate understandings, including mystical and religious interpretations (Davis 1998; Lieb 1998).

As with many areas of fringe and pseudoscience, there can be a proximity between science and other belief- or knowledge-systems, including the mystic or magical. Indeed, as Erik Davis (1998) shows throughout his brilliant book *TechGnosis*, there is an entire 'secret history' of technomysticism running alongside more orderly, rational stories of technology.[7] Each new technology brings with it new magic: the telegraph, radio and telephone all connect to spiritualism, to hearing the voices of the dead, for example. The New Age counterculture similarly made much of the links between technology and transcendence (or at least trance), notably in the work of shamans and gurus such as Timothy Leary (see Leary 1994). And certainly there is much evidence of magical ideas about cyberspace to be found out there, much of it overwritten by a technophilic utopianism in which the Internet is going to conjure higher states of collective consciousness. This is particularly acutely expressed in assorted 'high tech subcultures', such as the **Extropians**, who have built a new religion of self-improvement through technological enhancement (Terranova 2000).

But the counterculture and cyberspace have birthed other viewpoints on technology, too, including those that see nothing magical, in fact nothing positive, there at all. If the discussion above can be characterized as the public *misrecognizing* of science – at least in terms of how mainstreamers would label the knowledges and practices we have been looking it – what we move on to discuss now must surely be called the *public mistrusting of science* (and technology, of course). For as well as producing dreams of shiny new futures, scientific and technological 'advances' also provoke anxieties, phobias, panics. For every utopia promised, a parallel dystopia is threatened. The Neo-Luddites give a clear expression of this, rejecting not only computers, but (to varying degrees) other technological monstrosities such as the motorcar, electric lights – and even the switching of clocktime to daylight saving time (the US equivalent of the UK's switch from GMT to BST). Dismissed as crackpots by a lot of people, the Neo-Luddites are nevertheless mobilizing a more widespread set of fears about the effects of technology. While this might be brushed aside as naïve determinism, we've already seen in Chapter 3 that the pervasiveness of deterministic thinking, even in paranoid forms, means we can't afford to overlook its hold on understandings of science, technology and culture. In another attempt at symmetry, then, it is important to think about the work going on in species of anti-scientific and anti-technological thought and practice.

A less frightened, but no less deterministic variant on Neo-Luddism stresses that new sciences and new technologies bring new problems as much as they offer new solutions. Edward Tenner (1996) discusses this 'revenge effect' – antibiotics bring resistant strains of bugs; new buildings make us sick; traffic management causes jams, and so on. Or, as Paul Virilio (1989) puts it, before cars there were no car crashes, before nuclear power no Chernobyl. This leads him to propose a Museum of Accidents as an antidote to the safe stories told in science museums. So, according to this logic, without the ubiquity of the silicon chip, there would have been no Millennium Bug panic (Tapia 2003); or, to step back in time, without fire no one gets burnt.

Of course, one dominant strain of antiscience and technology emphasizes environmental impacts; versions of 'eco-criticism' spin this differently, but all are joined by a stress on science and technology's destructive consequences. From the Green Party to deep green 'monkey-wrenchers', from recyclers to lesbian separatists, environmentalists of all hues have mobilized around the truly Earth-shattering 'revenge effects' that our push for progress has wrought (Milton 1993; Redclift and Benton 1994). Of course, not all 'greens' are antiscience, and for many the answer to environmental problems is going to come from science, whether mainstream or fringe. But other strands are more rejectionist, refusing the idea of science as well as its artefacts and effects (or, perhaps, *because of* its artefacts and effects). Finding alternative, sustainable, organic ways of living has also moved from being a fringe pursuit into the mainstreams of popular culture and everyday life – though this fact annoys some scientists, who see irrationalism at work in public scepticism about genetically modified foods, for example.

Bruno Latour (1993) uses an environmental issue to illustrate his twin modern processes of purification and translation, in *We Have Never Been Modern* (see also Chapter 3): the hole in the ozone layer. As he shows, this is at once scientific and social, a problem of the atmosphere and of everyday life. Exactly such a reading has led some critics to bundle social studies of science and technology into the antiscience camp, too (see, for example, Williams 2000). This is, of course, what got misrecognized and misrepresented in the Science Wars – the idea that sociologists and cultural critics, and the broader 'academic left', is at heart against science, not least because it is a metanarrative. Dismissed as 'delegitimators' by defenders occupying an 'anti-antiscience' position (Segerstråle 2000), the Science Wars raging here have in fact been fought in a long-term war of attrition in the borderlands of science and technology. But who is patrolling those borders, and what are they armed with?

ScienceCop

Barry Barnes *et al.* (1996) write that boundary policing can take a number of forms. Alternative knowledges can be straightforwardly ignored – the case they give is Michel Gauquelin's (1983) statistical work on astrology, which showed that there were forces at work there that could not be explained by mainstream science. By sidelining such contrary evidence, the border guards are turning it into non-evidence, something not even worth contesting. As we've seen, a response to this cold shouldering in the fringes has been the adoption of rigorous scientific method, in order to force mainstream science to take notice. Such a stance is dealt with, Barnes *et al.* show, by ridicule and stigmatization.

Pseudoscientific discoveries and inventions are subject to extremes of nitpicking scientific scrutiny – so nitpicky, as Collins and Pinch (1982) argue, that much mainstream science would fail them. Scientists suggesting non-mainstream theories and findings have been professionally ostracized, publicly mocked, and even, if you believe the conspiracists, permanently silenced (Eisen 1999). Even when there's no whiff of a smoking gun, the effects of public pillory and excommunication can be professionally and personally devastating. Such discrediting can be meted out by individual scientists, by academic or professional bodies, or by self-appointed border patrollers such as CSICOP, the ironically sci-fi-like acronym of the Committee for the Scientific Investigation of Claims of the Paranormal, sketched by Andrew Ross (1994: 18) as 'an international "inquisition" of mostly academic ghostbusters, set up in the mid-seventies not only to combat the rise of the Christian fundamentalists' creationist claims, but also actively to police the boundary between science and pseudoscience'. Carl Sagan was a prominent member, as are several magicians, who use sleights of hand to replicate paranormal events. It publishes the debunkers' bible, *The Skeptical Inquirer*.

Ross reads CSICOP as a rearguard defence against the dark arts of New Agers, a reactionary rationalism whose very existence suggests that science is on the ropes. Sagan (1997), wanting to find some middle ground so that science doesn't seem too stuffy or sniffy, ends up patronizing the poor fools who misbelieve, urging 'a compassionate approach' in the hope of making science 'less off-putting, especially to the young' (282–3). During the Science Wars, CSICOP raged against the 'threat' of antiscience through the pages of *The Skeptical Inquirer*, which printed a top ten hit-list of causes of antiscience, here retold by Ullica Segerstråle:

1. The anxiety about a possible nuclear holocaust after World War II
2. Fears generated by the environmental movement
3. Widespread phobia about chemical additives
4. Suspicion of biogenetic engineering
5. Widespread attack on orthodox medicine
6. The growing opposition to psychiatry
7. The phenomenal growth in 'alternative health cures'
8. The impact of Asian mysticism in the form of Yoga and various spiritual cures
9. The revival of fundamentalist religion even in advanced scientific and educational societies; creationism in the United States
10. The growth of multicultural and feminist critiques of science education

(Segerstråle 2000: 80)

As Segerstråle reports, Gerald Holton's (1993) *Science and Anti-science* was excerpted in the same issue. Holton has his own list, of four types of delegitimator: philosophers and sociologists of science (Horton cites Latour), 'alienated intellectuals', New Agers, and radical feminists. Presumably those of us to whom one or more of these labels apply can also look forward to some Sagan-style 'compassion', and a helping hand back on to the one true path (or the one path to truth)!

But CSICOP operations have their revenge effects, too. As Jane Gregory (2003) shows in her exemplary account of the fate of astronomer Fred Hoyle, discussed earlier, the closing off of scientific channels for his ideas led Hoyle to take a populist route, in both factual and fictional science writing. Hounded out of the scientific mainstream, Hoyle used popular science channels to promote panspermia and to accuse the *Archaeopteryx* of faking it. While such a manoeuvre provides ammunition for those wanting to collapse pop science in with fringe and pseudoscience, it can also be read as a sign of popular culture's openness, marking the popular as an important site of a different kind of boundary work.

The trouble with no boundaries

So far my discussion has *tried* to be even-handed; though you may have detected a bit of siding with the underdog, something of that 'perverse taste for the margins'. My attempt at 'symmetrical thinking' is, of course, exactly the kind of thing that can land one in the hot water of 'postmodern relativism' – as you can see, for example, in Sagan's (1997) comments on Hess's (1993) *Science in the New Age*, which grumble about his uncritical discussion of various pseudo-sciences. And of course it dogged the Strong Programme too (see Chapter 2). But to return to my earlier comment, does having an inkling of a liking for Fred Hoyle mean I also have to agree that Dr Atkins has got it right? When people tell me that they've lost weight by going low on carbs, should I treat them with the same generosity (or more, or less) that I would treat anyone who says they've seen a UFO, or who avidly reads their daily horoscope? And, like countless parents caught in conflicting advice and emotions about MMR, who can I turn to, to guide me through this, or any other, maze?

One of the interesting but troubling things about margins is their very *equivalence*: once you wander right out there, far away from familiar landmarks, and the kinds of strangeness multiply and diversify, there comes a point where judgement gets fuzzy, where there are slippages between things, a kind of flatness. If I like TV bushcraft expert Ray Mears, does that mean I have survivalist tendencies?[8] Once we've strayed well away from orthodoxy, does that mean there can be *no limits* on the unorthodoxies we must accept? As Frank Trocco (1998) reminds us, the scientific fringes are also home to holocaust deniers and creationists, not to mention 'technopocalyptic' sects (Bozeman 1997) and some good old-fashioned con artists and snake oil sellers. Of course, this worrying is exactly what lets rationalists throw babies out with bathwater – for every New Ager they can throw back a holocaust denier. Even sociologists of science, those supposed demon conjurers of antiscience, make distinctions, as when Barnes *et al.* (1996) describe acupuncture and water divining as 'beyond the pale'. And creation science has been widely attacked or dismissed, though Hoyle once spoke up in its defence, again voicing his scepticism about Darwinism (Gregory 2003).

So when Jodi Dean (1997) writes about the postmodern fugivity of truth, does this too mean we have dismantled the apparatus of truth-tests irreversibly? That we have, indeed, spawned a monster – the monster of relativism? I don't think it does; while I agree with Dean's (1998) desire to suspend judgement about the truth of UFOs and aliens, or Davis's (1998) call to explore *how* fringe ideas work, rather than fixating on their veracity – and while I still want to privilege *The X-Files* mantra 'I want to believe' above CSICOP's 'We will not believe' – is this a good-enough position to hold? I think that Andrew Ross (1991) provides the best routemap, in his famous discussion of hacking. He calls on cultural critics to become hacker-like, to critique from within, to use 'technoliteracy' as a means to contest both mainstream and marginal science and technology:

> If there is a challenge here for cultural critics, it might be the commitment to making our knowledge about technoculture something like a hacker's knowledge, capable of penetrating existing systems of rationality that might otherwise be seen as infallible; a hacker's knowledge, capable of reskilling, and therefore of rewriting, the cultural programs and reprogramming the social values that make room for new technologies . . . [W]e cannot afford to give up what technoliteracy we have acquired in deference to the vulgar faith that tells us it is always contaminated by the toxin of instrumental rationality . . .
>
> (Ross 1991: 100)

Such a 'technocultural criticism' means turning scientific knowledge back in on itself, means being at once insider and outsider, means a truly symmetrical scepticism, rather than one already weighted in favour of orthodoxy. The Science Wars, and the never ending boundary work that we have worked through in this chapter, should be enough to convince us of the urgency of this task as central to the project of cultural studies.

Further reading

Bryson, B. (2004) *A Short History of Nearly Everything*. London: Black Swan.

Collins, H. and Pinch, T. (1982) *Frames of Meaning: The Social Construction of Extraordinary Science*. London: Routledge and Kegan Paul.

Davis, E. (1998) *TechGnosis: Myth, Magic and Mysticism in the Age of Information*. London: Serpent's Tail.

Dean, J. (1998) *Aliens in America: Conspiracy Cultures from Outerspace to Cyberspace*. Ithaca NY: Cornell University Press.

Eisen, J. (1999) *Suppressed Inventions and Other Discoveries*. New York: Avery.

Gieryn, T. (1999) *Cultural Boundaries of Science: Credibility on the Line*. Chicago: University of Chicago Press.

Gregory, J. (2003) The popularization and excommunication of Fred Hoyle's 'life-from-space' theory, *Public Understanding of Science*, 12(1): 25–46.

Hess, D. (1993) *Science in the New Age: The Paranormal, its Defenders and Debunkers, and American Culture*. Madison WI: University of Wisconsin Press.

Ross, A. (1991) *Strange Weather: Culture, Science and Technology in the Age of Limits*. London: Verso.

Sagan, C. (1997) *The Demon-haunted World: Science as a Candle in the Dark*. London: Headline.

Notes

1 I have toyed with and tried using the term 'technoscience' throughout, but abandoned it for vague reasons to do with writing and reading. I hope you can permit me this, and keep in mind that the boundary work I am discussing here refers to technology as well as science – or, perhaps I should say, it all refers to technoscience.

2 Hawking later wrote the foreword to *The Physics of Star Trek*, as well as appearing on both *Star Trek* and *The Simpsons* – a rare double feat for a scientist (see Higley 1997)!

3 My use of terms like 'mainstream', 'non-mainstream' and 'alternative' is hesitant and clumsy. Rather than litter the page with scare quotes, however, I will trust the reader to share my hesitation and ambivalence, to know that it's hard to find the right words sometimes, and that right-ish words have to suffice.

4 Named after American writer Hunter S. Thompson's gonzo journalism, gonzo science stages outsiderhood through a website (www.gonzoscience.com) covering topics including alternative archaeology and rogue geology.

5 I borrow this phrase from the now-defunct *Counterproductions Catalogue*, my variant on the *Whole Earth Catalog*, which used to sell through mail order an amazing range of fringe books and magazines, including many that dealt with science and technology.

6 My late father, a sceptical consumer of popular science, introduced me to Hoyle when he read *Evolution from Space* in its 'pulp science' edition.

7 Although this is strongly rejected by other critics, who see such mysticism as another example of the 'triumph of the irrational'; see Stivers 2001. I prefer Davis's tack, however, which is to sidestep such critique through an interest in how techno-mysticism *works*, rather than in poking at its holes.

8 An explanation in light of my earlier worries about the ephemerality (and localness) of popular culture: Ray Mears writes books and makes TV programmes about bushcraft – ways of living with nature, in the wild, using age-old skills and knowledge. At the time of writing he is on BBC TV, paddling a canoe around, meeting other bushcrafters.

7 | PAUSE AND REWIND

I want to wind things up here by thinking through a few last issues and ideas. They are purposefully mundane, or ordinary; as I hope I have shown throughout *Science, Technology and Culture*, there's a great deal that can be got from thinking about mundane 'science and technological stuff', to tweak a lovely phrase from Mike Michael (2003). The first was sparked by the email that I quoted at the start of the book. Remember that one thing it reminded lecturers like me about new, incoming undergrads was that

- Atari predates them, as do vinyl albums (except as an elite product used by professional DJs). The expression 'You sound like a broken record' means nothing to them.
- Most have never had a reason to own a record player.

Setting the Atari bit aside – though that in itself is a fascinating tale (and see Laing 2004 if you want to reminisce) – this mention of the vinyl record and the record player provoked in me the usual mix of nostalgia, curiosity and, I am reluctant to admit, something like the 'soft determinism' that I have been so guarded about in this book (and see Mackenzie and Wajcman 1999). Then I remembered Turo-Kimmo Lehtonen's (2003) beautiful article about the biographies of some technological stuff, which includes some very resonant comments about the disappearance of older technologies and the concomitant loss of technique that accompanies this. There's a clear link, too, to the brief mention of the audio cassette in Chapter 4. So I want to spend a few moments thinking about these issues, and why they did provoke in me the kind of deterministic thoughts that I should be deconstructing.

In a rich and nuanced discussion, making fantastic use of participants' accounts, Lehtonen sets out to explore, among other things, the relegation of older technologies as a newer one comes along. Like the displacement of Woody by Buzz Lightyear that frames the movie *Toy Story*, which allegorizes the supplanting of 'hand-made' cartoons by those produced digitally – this is kind of a sad story; sad in the way it shows how the churn rate of technological change renders older things obsolete so easily. Yet something 'sticks', too. The movement from one technological solution to another isn't so smooth; and most of this stickiness is cultural. It's about the attachments and habits that build up around ways of doing things with technology, and it cuts through the smooth logic of technological progress, showing that ideas like functionality or usability aren't always on an upward curve. It shows that people aren't simply being technophobic when they refuse to adapt readily to the life changes that a new technology demands of them; and it reminds us that thinking about culture as 'ways of life' opens up for enquiry a whole set of force-fields, with different intensities and directions. Lehtonen poses, then answers, the key question I am trying to move towards:

> What does it mean for a technology to become 'old'? In fact, when things are not actually broken, age in itself does not appear to be as important as the decline in the amount of use and the feeling that the device does not fit in anywhere anymore.
>
> (Lehtonen 2003: 378)

One of the examples Lehtonen uses to flesh this out is music technology – specifically vinyl and audio cassette technologies, both of which have been pushed aside by the compact disc (CD). Or, as he rightly says, the vinyl record and its playing machinery have become the province of specialists, who have remade a once very mundane set of practices into a kind of retro-techno craft skill.

Collectors and nostalgists have similarly revalued what had been devalued by technological newness, in a version of the strange mix of 'hipness and nerdiness' that Will Straw (1997: 9) says characterizes popular music collectors (see Higgins 2001; Mulholland 2002 for different expressions of this). The same, of course, is true for other technologies with high churn rates, such as computers (Finn 2001; Laing 2004). Even outside the world of collectors, Lehtonen notes that 'people have a hard time abandoning their unused devices once and for all' (379). And I have to admit that I was as recalcitrant as a scallop when CDs first came on to the market. Part of this is, of course, eco-nomic. I had lots of records (and tapes) that I didn't want to 'abandon', or have to replace with miniature facsimiles. I didn't want to be cajoled by the record industry, even as vinyl and tape started to vanish from shops (apart from specialists, of course). But it's also partly to do with technique, in the sense of embodied, often habituated knowledge and skill (which in being habituated ceases to be seen as knowledge or skill at all). Lehtonen is spot on about this, too: 'once a new standard of technology takes over, and old things are abandoned, the old body of knowledge will be transformed, if not lost. Most people under 25', he adds, 'barely know how to use a vinyl record player anymore' (382). He thus concurs with Michael's (2003) useful discussion of the role of mundane technologies in 'disordering' – his argument that it isn't just new, 'exotic' technologies that produce change, but also older, more mundane ones, too (contrary to the assumption that their main role is social reproduction or ordering).

Lehtonen's paper also makes me rethink those lost skills, and their manifold pleasures (but also problems). The smell of records, the visual feast of a gatefold sleeve – and the way that carrying a 12-inch record offered opportunities for visual display of music connoisseurship. Then the act of playing a record, how to hold the record by its edges, blow the dust off, cue up the arm, gently land the needle in the groove (having first set the correct RPM speed on the turntable). But also the scratches, the warping, the jumping, the crackle (there was a fad for faking this crackle more recently, on CDs; we should also note that CDs have brought their own techniques, as well as their own problems, despite the early sales-pitch about their indestructibility). Even without the newer techno-skills of scratching that hip hop brought to the (turn) table (see Bennett 2001 for a run-through), what we have here, I think, is a lovely example of something Donna Haraway (1991: 180) talked about, albeit in a very different context and with a very different purpose: 'intense pleasure in skill, machine skill, [as] an aspect of embodiment'.

Now, the ambivalence that shades this discussion, as already apologized for, is its slide towards 'soft determinism'. The shift from vinyl and tape to CD did make me feel *subject to* technology; I did feel that my machine skill had been devalued, that I had been deskilled. I wanted to resist, even as resistance became futile. Then I surrendered, slowly entered 'CD culture', reshaped my life's daily patterns around this, invested in its hardware and software, learnt the new skills and made all the other necessary adjustments – things like storage, and also, to echo Lehtonen, the partial abandonment of the old (but not its complete disposal). I think this can help us understand why technological determinism has such popular purchase: because at an individualized level – the scale at which we are encouraged to think about things, let's be honest – the traffic does seem overwhelmingly one way. I didn't shape CD technology as a recalcitrant user, other than by joining unknowingly with other 'laggards' and necessitating further efforts on the part of manufacturers and retailers to convince or cajole us into switching (we can see the same things going on at the moment around televisions, with bewildering new forms of flat screen, LCD, hi-definition, etc.

vying for our attention and money). But, you see, even by thinking about all the people like me who sulked about CDs for a time shows some two-way traffic, a feeding back of recalcitrance, reflected in sales figures and buyer profiles, that prompted renewed effort to 'sell' the CD to us. That's why Lehtonen is dead right in his formulation of the double meaning of 'trial' in the domestication of technology, and his reminder that domestication means 'living with', implies change on both sides, reciprocity. Seeing this requires thinking along the lines of those force-fields I mentioned earlier, and acknowledging the unpredictable plot twists of the stories of science, technology and culture.

I called this chapter 'Pause and Rewind' not just because that's what I want you to do, but also because those two words make me think about the ideas of functionality and usability anew. At the moment, Sky TV is heavily hyping its Sky+ technology, which allows viewers apparently to pause and rewind live television (see O'Sullivan 2005). The advert I keep seeing is both seductive and frustrating, presumably on purpose: by pausing the image of the BASE-jumper before his or her parachute opens, then rewinding back to the moment of jumping, it suggests that we can wrest the control of TV images away from the broadcaster, who here controls the action, and become its masters our-selves. Seeing this advert for the thousandth time set me thinking about those twin functions of assorted recording and playback technologies – *pause* (the ability to temporarily stop the flow of image and/or sound at a particular point, and then resume at that exact point) and *rewind* (the function of moving quickly backward through a recording, either back to the beginning or to a particular spot). There is also, of course, rewind's opposing function, fast forward. Digital recording technologies, whether audio or audiovisual, seem to have improved the functionality of pausing: video tape in particular is prone to flickering when paused, while the mechanical, motorized working of tape play-back means that pausing always puts a strain on both tape and motor. Furthermore, the mechanics of tape machines mean that pausing and 'unpausing', as a mechanical process, involves a small time delay, as the motor gets the tape moving again (endlessly frustrating for the home recordist). With CD and DVD, pausing stops and then restarts things dead, dead still, and instantly. And rewinding has all-but vanished from digital storage and retrieval; there's no need to wind a DVD back to the beginning once you've watched it. (But then this also means that you can't, so far as I can see at least, stop a DVD at a particular point, take it from one player to another, and restart it at that point.) The art of good rewinding is becoming lost knowledge and skill, and I can only begin to ponder the losses and gains that have resulted. Example: gain – no tape strain from repeated winding back; loss – the moments of quite contemplation and anticipation as the video rewinds.

OK, so the gains do seem to outweigh the losses here, and the loss I have pointed out seems quite trivial. But in a way it isn't, because it signals something bigger, to do with time and our experience of technology (see Michael 2003 for a different take on technology and temporality). One of the structuring logics of technological progress is increasing speed (others include increasing smallness, user-friendliness, etc.). We don't want to have to wait for a tape to rewind, we want instant playback! It's the same when you switch on your computer, and get twitchy waiting for it to boot up. Come on, faster! I haven't got all day! Faster means better (see Eriksen 2001 on our 'hurried era').

Now, mindful that I am once more veering towards 'soft determinism', I do think there is a serious point here about speed and slowness. As Wendy Parkins and Geoffrey Craig (2005) show in their book *Slow Living*, there is a growing counter-movement that wants us to resist the acceleration of everyday life, that asks that we reappraise the cultural value of 'quick' and 'slow'. While this isn't only about technology, there's clearly a 'hard determinism' at work in many of the slow livers' pronouncements. Having to rewind a tape, and wait while this happens, isn't that unlike *Slow Living*'s accounts of the pleasures of taking all day to make dinner as opposed to 'nuking' a ready meal straight from the freezer (on the ambivalence towards microwaves and the idea of 'proper' cooking, see also Ormrod 1994). So a change which is supposed straightforwardly to enhance the usability of recording and retrieval technologies connects out to broader cultural valuations of temporality (among other things).

This discussion, although a bit fuzzy perhaps, suggests a need to 'think different' at every moment of considering the narratives around technology. The narrative of usability – which is supposed to be uncontested, as surely everyone wants things to get easier to use – has to be run alongside those ideas about 'machine skill', which can mean pleasure in usability, as well as worry over deskilling or lost knowledges. A vivid example that we can turn to briefly here concerns the now-ubiquitous 'presentational software' package, Microsoft PowerPoint.

Giving good PowerPoint[1]

Sherry Turkle (2003) has written a brief, insightful paper about the use of PowerPoint in schools in the USA, observing how it has transformed presenting and also, in her view, arguing and thinking:

> The software does not encourage students to make an argument. They are encouraged to make a point. PowerPoint encourages presentation not

conversation. Students grow unaccustomed to being challenged. Ambiguity is not valued. A strong presentation closes down debate rather than opening it up because it coveys absolute authority.

(Turkle 2003: 23)

In a neatly telling phrase, Turkle refers to PowerPoint's 'special effects' (24) – all the animations, sounds, whiz-bangs of this 'new presentational aesthetic', that get factored in to making a presentation full of awe – but awe at the aesthetic rather than the content. The content, the ideas, Turkle says, are squished down to bullet points, so that PowerPoint makes us tell simple, tidy stories – what Coleman (2003) refers to as 'the homogenization of public discourse at meetings' (and at conferences, in the classroom, etc.). Critics suggest that the presentational aesthetic of PowerPoint overshadows a more discursive approach to learning or sharing ideas. A polished set of slides makes the ideas seem equally polished, finished, closing down possibilities for change and revision. Moreover, like CGI special effects in cinema (see Chapter 4), this new aesthetic announces itself in terms of technical mastery first – giving good PowerPoint. Now, while Turkle's article could be drawn into the growing mass of anti-PowerPoint rants and arguments currently floating round cyberspace,[2] it's important to see that this isn't a simple case of **technophobia**.

I think the anti-PowerPoint issue is nicely resonant of bigger issues at stake here, in terms of my founding question: what does it mean to think about science and technology as culture? It means examining the cultural work that PowerPoint does, and the cultural work that made PowerPoint take the shape it does. Perhaps Matthew Fuller (2003) said it best, in his call for 'software criticism', exemplified in his detailed discussion of Microsoft Word and its relation to (note: *not just impact on*) computing, writing and thinking. While Fuller is 'against' Word in the same way Turkle is 'against' PowerPoint, attention to the assembling of arguments without closing off conflicting viewpoints (exactly the kind of thing that Turkles says PowerPoint erases) allows us to 'think different', again, about the interleaving of science, technology and culture.

In the end the crucial point – and this is, as Keller (2003: 5) puts it, a 'verbal point, not a PowerPoint point' – is about the tools we use to think about the tools we use. And it's here, I think, that cultural studies of science and technology has so much to offer, with its scruffy mixture of theories and methods, and with its political commitment to understanding both the promises and threats of science and technology, refusing any one-size-fits-all perspective (see Balsamo 1998; Best and Kellner 2001 for more on this). So, before I finally wrap things up, I want to turn briefly to one final area of current fascination and anxiety in contemporary technoscientific culture: cosmetic surgery.

Nipped and tucked

There's no denying that cosmetic surgery offers a potent lens for considering the ambivalences of science, technology and culture. Just off our screens after a controversial run, the TV show *Cosmetic Surgery Live* brought just about every response imaginable out of its fans and critics – and those who refused to watch such 'lurid' programming. On a different channel, the hit US show *Nip/Tuck* dramatizes the life and work of two cosmetic surgeons, bringing many of the same issues and reactions to the surface. *Cosmetic Surgery Live*, as its name blankly suggests, provided viewers with real-time footage of surgical procedures, as well as offering advice for those considering surgery (viewers could video-text pictures of the body-parts in question and get an instant response from the show's experts). Its liveness heightened the speed of surgery, but invisibilized things like recovery and aftercare; it also foregrounded technologies not only of surgery but also of media and communications, with live feeds of procedures from surgeries on both sides of the Atlantic, not to mention the video-texting noted already. The programme at once normalized cosmetic surgery by making it seem so easy and quick, and also sensationalized it by showing graphic detail. Like previous live anatomical-science spectacles, it revealed 'the body emblazoned', its skin flayed to reveal the flesh and fat, the blood and guts (Foucault 1975). It offered a 'poor man's visible human project' (see Chapter 1), with the added bonus of another form of liveness: the liveliness of the patients, who were often conscious during procedures, offering their own 'warm expertise' on the process.

Outside of televisual contexts like *Cosmetic Surgery Live*, the phenomeon has received considerable academic scrutiny, particularly in terms of its 'normalizing' effects on the gendered and racialized body (Davis 1995, 2003; Morgan 2003). Such analysis has two threads to it, which often rub up against one another, as Ruth Holliday (2005b) has shown. One is the quest to be 'normal', the other the desire to be 'beautiful' – of course, both of these are immensely loaded notions. The rapid consumerization of cosmetic surgery – vividly dramatized in *Nip/Tuck* through the surgeons' consulting question 'Tell us what you don't like about yourself' and through the shopping-like behaviour of their clients – Holliday notes, suggests a proliferation of *difference* as much as a narrowing of normal. In her own follow-up work, with Jackie Sanchez Taylor, this idea is given more flesh, in an attempt to read cosmetic surgery in the context of contemporary theories of reflexive self-identity (Holliday and Sanchez Taylor 2006). Such an argument is one way of countering technological determinism, as we have seen throughout this book. It situates the self within a network, a force-field, that has conflicting tendencies, alternating currents, competing agendas. It refuses to dismiss lines of thinking without having first thought them through, and even then it retains an interest in the cultural work done by lines of thinking it doesn't find useful. So, instead of Morgan's (2003: 176) question 'Are there any politically correct feminist responses to cosmetic surgery?' (the answer seems as if it's going to be 'Maybe', but actually ends up being 'No'), we have to look instead at the range of responses, experiences and representations, the sites where knowledge and practice are produced and consumed, the discourses and identities that feed into and spin out of a techno-scientific and cultural phenomenon such as cosmetic surgery (and, indeed, PowerPoint). The task becomes much bigger and more complex, to be sure, but then these are big, complex things. They're also, as Balsamo (1998: 299) writes, 'exciting and cranky'. Like big science, it seems we need 'big culture' to help us work our way through all this, 'big culture' as a field of study of science and technology that is exciting, but also proud of being cranky. And just like the work of geology described by Raab and Frodeman (2002: 76), such a project means viewing science, technology and culture 'from moving and shifting perspectives where a variety of scales intersect and play off one another'. My task in this book has been to map some of those scales and perspectives, with the overall aim of showing what it means to think of science and technology as culture. This has indeed meant playing with a variety of scales and perspectives, in an effort to explore, among other things, what happens when we approach scientific practice as cultural practice, or when we try to read technological artefacts as cultural artefacts. It has meant looking in more detail at some manifestations of technoscientific culture, such as Moon landings and atom bombs, and it has meant poking around in the misty margins that supposedly demark the 'edges' of scientific knowledge. As a still-emerging, still-growing field of enquiry, cultural studies of science and technology can no doubt look forward to more poking around, more pondering, and more controversy. We wouldn't have it any other way.

Notes

1 This phrase is adapted from Turkle's (2003: 23) note that school teachers take books off reading lists if those books 'don't give good PowerPoint', meaning that they don't present their information in a way that lends itself to instant transfer to a slideshow. In fact, textbooks are marketed today with ready-made PowerPoint accessories, negating even the act of transfer, let alone any work of translation.

2 Typing key words such as 'hate', 'loathe', 'despise' and 'PowerPoint' into Google brings up a rich seam of websites, blogs and so on. Some of it is fairly puerile, and there's a lot of repetition of the 'PowerPoint destroys thought' variety, but also some interesting viewpoints.

GLOSSARY OF KEY TERMS

Actor-network theory (ANT): Offshoot of the sociology of scientific knowledge, associated with the work of, among others, Michel Callon and Bruno Latour. ANT seeks to understand how the practices of science or technology revolve around the progressive constitution of a network made up of both human and non-human actors. It refuses to differentiate different types of actant, and focuses on the processes by which actants are enrolled into the network in order to produce desired outcomes. ANT has produced some notable studies of particular networks, for example Callon (1986) and Latour (1992), but has also been continually critiqued, even by those theorists most closely associated with it (see Law and Hassard 1999).

Apollo: A central symbol of the Cold War space race, NASA's Apollo programme (1963–72) was devised to land (American) humans on the Moon and bring them safely back to Earth, and six of its missions (Apollos 11, 12, 14, 15, 16 and 17) achieved this. (Apollos 7 and 9 were Earth orbiting missions to test the Command and Lunar Modules, and did not return lunar data. Apollos 8 and 10 tested various components while orbiting the Moon, and returned photography of the lunar surface. Apollo 13 did not land on the Moon due to a malfunction, later made into a famous movie.) The six missions that landed on the Moon returned a wealth of scientific data, 400 kilogrammes of Moon rock, and some iconic photographs (Light 1999). Three more lunar missions were planned (Apollos 18 to 20), but the reduced NASA budget meant that these missions were cancelled to make funds available for the development of the Space Shuttle and the Skylab programme. See http://www.nasm.si.edu/collections/imagery/apollo/apollo.htm

ASCII (American Standard Code for Information Interchange): A standard character set and character encoding based on the Roman alphabet most commonly used by computers and other communication equipment to represent text. Particularly associated with dot-matrix printing, the ASCII aesthetic is nowadays seen as tech-nostalgic, and is used, for example, to produce distinctive artworks in which images are converted to ASCII text.

Atkins diet: Weight loss dietary regime popularized by Dr Robert Atkins which equates weight loss with the control of carbohydrate intake, rather than calories or fats. It is seen by many medics and nutritionists as contradicting scientific wisdom about the connections between diet and body weight. The low intake of carbohydrate produces a condition called ketosis, which is key to the Atkins diet and is believed to encourage the body to burn off its fat reservoirs. Hugely popular and widely regarded as a successful weight-loss diet, the Atkins diet (and other low-carb diets) remain outside of nutritional orthodoxy, and are even considered to be detrimental to the health of participants.

Big science: Term used by scientists and historians of science to describe the change in science which occurred in industrial nations after World War II. The development of atomic weapons during the War pointed to the centrality of science to political power, so governments became the chief patrons of science, and the character of the scientific establishment underwent several key changes. The bigness of big science refers to its budgets, the number of scientists involved in particular projects, the scale of equipment used and the centralization of scientific practice into large labs and institutions. Big science is controversial, for example because it is seen to undermine scientific method, because it is elitist, because its weds science to funders such as governments or the military, and because it has led to the bureaucratization of science. See Sassower (1995).

Black box: A simple input-output device. To black box something means to make its inner workings invisible. Technology is black boxed for the majority of its users: we don't know how it works, or even what's inside it, but we know just enough to make it work. Black boxing can be seen as a way to control scientific or technological knowledge, by making it seem too difficult for 'ordinary people' to understand. See Sismondo (2004).

Complementary and alternative medicine (CAM): A diverse group of medical and health care practices, systems and products that are not presently considered to be part of conventional or orthodox medicine. There have been long-running debates in the West over the status of these forms of medicine, which are battles over the legitimacy of different understandings of health, illness and treatment. The list of what is considered to be CAM changes continually, as those therapies that are proven to be safe and effective become adopted into conventional health care, while new forms of CAM are introduced. Complementary medicine is used together with conventional medicine, whereas alternative medicine is used in place of conventional medicine. See Saks (2003).

Conspiracy theory: Countercultural systems of belief that suggest some form of 'cover-up', usually by the government (sometimes in collusion with other agencies, both human and non-human). Prominent conspiracies include the cover-up of US government possession of UFOs and extraterrestrial technology, or those connected to the deaths of prominent public figures (President Kennedy, Marilyn Monroe, Princess Diana). There is a dense intertexual web of conspiracy theories in circulation, even more so thanks to the Internet (see Thieme 2000). Conspiracy theories can be seen as forms of contested knowledge that have their own logic and structure (e.g. that lack of evidence serves to prove a cover-up). See Knight (2000).

Delegation: As it is used in discussions of technology, delegation refers to the use of technological artefacts to do the work of humans; this work often involves ways of 'controlling' the behaviour of people (see Latour 1992). Human societies, Latour argues, are held in place through the delegation of many tasks to non-humans, such as tools or texts.

Double life of technology: The idea that technological artefacts can be put to uses other than those they were designed and devised for. Inventors, designers and manufacturers may attempt to fix the use (and meaning) of a new device, but once it is being used, its double life may emerge as users adapt its functions.

Downwinders: Term used to describe people who live or work in the vicinity of nuclear testing, particularly down wind of test sites, and whose health has been affected by airborne fallout. Seen as having been ignored by the government and military, and as victims of nuclear war. See Davis (1993).

Empiricism: The idea, central to the modern scientific method, that theories should be based on observations of the world. Recent developments such as quantum mechanics are seen to challenge empiricism as the exclusive way in which science works and should work, because they are unobservable or purely theoretical.

Extropians: Followers of the principals of Extropy, which began in the late 1980s to outline an approach to the future that would maximize the positive benefits of new scientific and technological developments. The principles of Extropy are Perpetual Progress, Self-Transformation, Practical Optimism, Intelligent Technology, Open Society, Self-Direction and Rational Thinking. Described by Terranova (2000) as one of a growing number of 'high-tech subcultures', Extropians share with other groups a faith in a posthuman future. See http://www.extropy.org/principles.htm

Hacking: Commonly associated with malicious or criminal activities, especially over the Internet, hacking has become a prominent site of anxiety around computers and cyberspace. While its origins in the computing subculture show that it should be thought of more broadly as a way of displaying talent with programming, it is overwhelmingly associated with cybercrime, covering a variety of activities that breach online security systems, either to access confidential information or for financial gain. The writing and sending of computer viruses has become a prominent form of hacking. See Ross (1991); Taylor (1999).

Heterogeneous engineering: Term used in ANT to describe how actors assemble networks by drawing together very different people and things. See Law (1987).

Internet: The global network of networks that connects millions of computers and facilitates things like email and browsing the web. Developed in the 1960s, initially as the ARPANet, the Internet has a history tied to the military-industrial-entertainment complex, well covered in Abbate (2000). Still rapidly expanding and being put to new uses, the Internet has also become a taken-for-granted technology of modern living for many people, though it has also introduced a new way for society to be stratified, the so-called digital divide. It has also generated endless debate, from every conceivable angle.

Intertextuality: Concept in literary theory that describes the connections between 'texts' (used in the broad sense to mean any cultural products, such as books, films, songs, etc). Intertextuality can mean the way that, say, a film-maker 'references' other films in her or his work, or the ways that people 'reading' the text make those connections.

Luddism: The Luddites were nineteenth-century activists who protested against changes to the pattern of work, especially due to its mechanization, most famously by smashing the machines that they saw as deskilling them and robbing them of work. The term Luddism is (wrongly, some people argue) therefore used more broadly to describe a form of radical technophobia; so-called New Luddites also stage machine-breaking, only they smash computers. For a critical discussion, see Robins and Webster (1999).

MMR (mumps, measles and rubella vaccine): Controversial triple vaccine used to prevent three diseases, administered routinely to babies. A link between the vaccine and subsequent health problems, such as the onset of autism, prompted a health scare in the UK, and a widespread popular debate about not only the vaccine, but the relations between the government, the health service and pharmaceutical companies. Many parents opted not to let their children be vaccinated, but the government continued to refute the findings that connected the MMR vaccine to health problems. The MMR issue highlights the trust/risk balance in modern life, and also the ways in which medical science polices its boundaries, often with assistance from the state.

Nanoscience: The emerging science of very small things, which brings together chemistry, physics, biology in a new science. The word 'nano' means 10^{-9} – a human hair is about 100,000 nanometres thick. At the nano scale, materials have very different properties, which can be exploited to produce new nanotechnologies – tiny machines with a host of potential uses, such as in medicine.

Naturecultures: Donna Haraway (2003) uses this neologism to emphasize the impossibility of separating the natural and the cultural – what we think of as 'nature' is itself a product of culture. In her work on 'companion species', Haraway talks of the naturecultures of the co-evolution of dogs and humans, for example.

Nuclearism: Defined by Irwin *et al.* (2000: 90) as 'the popular negotiation of the ideological relations embedded in discourses about nuclear weapons as they work to authorize a continued need for their threatened use', nuclearism describes how the idea of nuclear weapons has been normalized, for example through the notion of mutually assured destruction (MAD), or nuclear deterrence: the notion that to have nuclear arms means to prevent their use.

Phenomenology: The philosophical study of experience, or consciousness: the appearances of things, or things as they appear in our experience, or the ways we experience things, thus the meanings things have in our experience. Phenomenology studies conscious experience as from the subjective or first person point of view. Key thinkers include Edmund Husserl, Martin Heidegger, Maurice Merleau-Ponty and Jean-Paul Sartre. Phenomenology addresses the meaning things have in our experience – the significance of objects, events, time, people, as these things are experienced in our 'lifeworld'.

Plate tectonics: Widely accepted theory about the structure of the Earth, which states that the crust of the planet is made of moving plates, and that there are places where the plates are formed, such as mid-oceanic ridges, and other places where they sink under one another, such as deep sea trenches. The 'engine' for plate movement is believed to be radioactivity-generated heat in the Earth's mantle. Alfred Wegener (1880–1930) first proposed that the continents were once compressed into a single proto-continent which he called Pangaea, and over time they have drifted apart into their current distribution. This continental drift idea helped account for the distribution of fossil organisms and the formation of major landforms such as mountain ranges. In 1929, Arthur Holmes suggested thermal convection as the mechanism causing drift. Both ideas were largely dismissed until the 1960s, when oceanographers found ridges and trenches, supporting the idea of seafloor spreading. There are believed to be ten major plates: the African, Antarctic, Australian, Eurasian, North American, South American, Pacific, Cocos, Nazca and Indian plates. Plate tectonics has moved from being a contested, 'fringe' theory in geology, to becoming very widely accepted within the scientific community.

Posthuman: The increasing interrelation of humans and technologies leads some thinkers to suggest that the very idea of being human, as something separate from being non-human or being technological, has ceased to make sense. We have become, in this argument, posthuman – a combination of human and technological which is inseparable (Badmington 2000). This idea is also discussed using the figure of the cyborg (see Haraway 1991).

Public understanding of science (PUS): The study of how non-scientists, or 'ordinary people', make sense of science. PUS is concerned with improving scientific communication, and often also with debunking pseudoscience. It also involves looking at science education, and trying to improve access to appropriately packaged scientific information. Popular science, for example television documentaries, is seen by some critics to improve PUS by making science accessible, while others see this as 'dumbing down'. Improving the public understanding of science is seen as vital to the continued legitimation of science, and to the policing of the boundaries of scientific knowledge. See Gregory and Miller (1998).

Purification: In *We Have Never Been Modern*, Bruno Latour (1993) argues that modernity has produced two contradictory impulses: one is making things simple, which he calls purification. The other, which actually involved making things more complicated by mixing different things together, he calls translation. Purification works by simplifying and classifying: the tidiest system has just two classes, which are mutually exclusive but which between them capture all possibilities. But the practices of modernity, such as science, actually produce things which defy this simple classification. The posthuman, for example, blurs the distinction between the human and the technological.

Realism: An approach to understanding how science 'works' in society, realism explores how things like scientific laws operate: a realist believes that science describes a 'real world', and moreover that this real world shapes science. The world must be a certain way that makes science possible. So realism argues that the objects of scientific knowledge are 'real' and exist outside of science, and science can tell us about this real world. This claim helps to justify the special status of science. See Yearley (2005).

Science Wars: Talking of science and technology as social or cultural things can be immensely unpopular – it can be seen as an attack on the legitimacy of science, for example. As social and cultural studies of science gained some prominence, so they encountered a backlash, notably from scientists. These so-called 'Science Wars' included the infamous Sokal incident, in which a scientist published an article in a social science journal, purportedly advancing a cultural studies of science standpoint, but later exposing himself as a prankster: the science used in the article was nonsense. This, Sokal said, proved that cultural studies of science doesn't understand science, and is a fruitless, pretentious exercise. See Ross (1996).

SETI: The Search for Extraterrestrial Intelligence, an exploratory science that seeks evidence of life in the universe by looking for some signature of its technology. SETI listens for radio noise coming from a distant planet, as a sign that life has developed there. Data collected by SETI is currently being sent over the Internet for processing by thousands of computers around the world, utilizing their idle processing time – this is the SETI@home project. See Squeri (2004).

Smart house: Rapid developments in assorted information, communications and media technologies, it has been predicted, will revolutionize our everyday lives. Nowhere is this prediction more vividly captured than in the idea of the smart house – a house in which integrated technology can intelligently manage many of the day-to-day activities of domestic life. A fridge that knows which food is nearing its sell-by date, and that can send a text message suggesting a suitable menu to use that food, or a room that senses your mood and can adjust the lighting and the muzak to suit you, are among the forms of smartness that future houses are predicted as offering their inhabitants.

Social construction of technology (SCOT): Branch of the sociology of science and technology, SCOT draws on work done in the sociology of scientific knowledge. SCOT provides an argument against technological determinism by showing how human action shapes technology, too. The 'shape' of technological artifacts is the outcome of social processes, such as the infuence of relevant social groups (for example, users).

Sociology of scientific knowledge (SSK): The application of sociological theories to understand the production, dissemination and consumption of science as a form of knowledge. Rather than seeing science as outside society or the social, SSK seeks to reveal the imprint of the social on scientific knowledge. It shows that science is part of society, and that science is shaped by social forces.

Strong Programme: A variety of the Sociology of Scientific Knowledge (SSK), the Strong Programme proposed that both 'good' and 'bad' science should be treated equally, or symmetrically – both are exposed to social factors like cultural context. A better understanding of the workings of SSK can be achieved by looking at the ways that different forms of knowledge are socially produced and consumed: all scientific knowledge is social, and what is 'true' is always contingent: it is better described as what is 'believed to be true'.

Teacher in Space Program: In an effort to better connect the American public with its Shuttle programme, NASA had the idea of sending a civilian into space, and in 1984, it invited America's teachers to submit applications for the Teacher in Space Program. Christa McAuliffe, a high school social studies teacher from Concord, New Hampshire, was selected to become a member of the crew of the Space Shuttle Challenger. Her role would be to teach lessons from the Shuttle to America's (and the world's) classrooms, and to share with them the experience of space flight – to re-enchant space flight in the public's eyes. Those lessons never occurred, because Challenger exploded soon after take-off, on 28 January 1986, while millions of schoolchildren looked on. See Penley (1997).

Technologicalness: A clumsy neologism to describe the extent to which technological artefacts 'announce' themselves (or are announced) as technological. As technologies become domesticated into our everyday lives, their technologicalness diminishes or recedes. Black boxing can be seen as a way to reduce technologicalness.

Technologies of the self: Term used by the French theorist Michel Foucault (2000) to describe a series of techniques through which individuals to work on themselves by regulating their bodies, their thoughts and their conduct. Technologies here refer to rules of conduct through which power is subtly exercised.

Technophobia: Used to describe the fear, or hatred, or suspicion of technology. By using the suffix '-phobia', the term suggests an irrational fear; critics would argue we are dead right to be fearful of technology.

Technoscience: A concept widely used in the interdisciplinary community of science and technology studies to designate the social and technological context of science. It is used to acknowledge that science and technology are inseparable, and that both are also inseparably social.

Technostalgia: I use this term to refer to a kind of nostalgic perspective on technology; collectors of retro computers, or ASCII artists, are enacting it. Technostalgia is about the ways in which, in the past, the future was imagined. See Taylor (2001) for a related discussion of technostalgic music.

Translation: See purification.

REFERENCES

Abbate, J. (2000) *Inventing the Internet*. Cambridge MA: MIT Press.

Adorno, T. (1994) *The Stars Down to Earth and Other Essays on the Irrational in Culture*. London: Routledge.

Allan, S. (2002) *Media, Risk and Science*. Buckingham: Open University Press.

Amis, M. (1987) *Einstein's Monsters*. London: Jonathan Cape.

Anderson, B. and Tracey, K. (2001) Digital living: the impact (or otherwise) of the Internet on everyday life, *American Behavioral Scientist*, 45(3): 456–75.

Appadurai, A. (1986) Introduction: commodities and the politics of value, in A. Appadurai (ed.) *The Social Life of Things: Commodities in Cultural Perspective*. Cambridge: Cambridge University Press.

Appleyard, B. (2003) Landing of hope and glory, *Sunday Times Magazine*, 14 December: 32–8.

Atkins, R. (2003) *Dr Atkins New Diet Cookbook*. London: Vermillion.

Badmington, N. (ed.) (2000) *Posthumanism*. Basingstoke: Palgrave.

Bakardjieva, M. (2005) *Internet Society: The Internet in Everyday Life*. London: Sage.

Bakardjieva, M. and Smith, R. (2001) The internet in everyday life: computer networking from the standpoint of the domestic user, *New Media & Society*, 3(1): 67–83.

Balsamo, A. (1998) Introduction, *Cultural Studies*, 12(3): 285–99.

Barnes, B., Bloor, D. and Henry, J. (1996) *Scientific Knowledge: A Sociological Analysis*. London: Athlone.

Bauman, Z. (1991) *Modernity and Ambivalence*. Cambridge: Polity.

Bauman, Z. (1997) *Postmodernity and its Discontents*. Cambridge: Polity.

Beck, U. (1992) *The Risk Society: Towards a New Modernity*. London: Sage.

Bell, D. (1999a) Bruno Latour, in E. Cashmore and C. Rojek (eds) *The Dictionary of Cultural Theorists*. London: Arnold.

Bell, D. (1999b) Secret science [review of Eisen 1999], *Science and Public Policy*, 26(6): 450.

Bell, D. (2001) *An Introduction to Cybercultures*. London: Routledge.

Bell, D. (2004a) AI (movie), in D. Bell, B. Loader, N. Pleace and D. Schuler, *Cyberculture: The Key Concepts*. London: Routledge.

Bell, D. (2004b) Memory, in D. Bell, B. Loader, N. Pleace and D. Schuler, *Cyberculture: the Key Concepts*. London: Routledge.

Bell, D. (2005) The culture(s) of cyberculture, paper presented at the Third International Conference on Communication and Reality, University Ramon Llull, Barcelona, May.

Bell, D. and Bennion-Nixon, L. (2001) The popular culture of conspiracy/the conspiracy of popular culture, in J. Parish and M. Parker (eds) *The Age of Anxiety: Conspiracy Theory and the Human Sciences*. Oxford: Blackwell.

Bell, D. and Binnie, J. (2000) *The Sexual Citizen: Queer Politics and Beyond*. Cambridge: Polity.

Bell, D. and Hollows, J. (eds) (2005) *Ordinary Lifestyles: Popular Media, Consumption and Taste*. Buckingham: Open University Press.

Benjamin, M. (2003) *Rocket Dreams*. London: Chatto & Windus.

Bennett, A. (2001) *Cultures of Popular Music*. Buckingham: Open University Press.

Benson, K. (2001) The Unabomber and the history of science, *Science, Technology & Human Values*, 26(1): 101–5.

Berg, A-J. (1994) A gendered socio-technical construction: the smart house, in C. Cockburn and R. Fürst-Dilić (eds) *Bringing Technology Home: Gender and Technology in a Changing Europe*. Buckingham: Open University Press.

Bertens, H. (1995) *The Idea of the Postmodern: A History*. London: Routledge.

Best, S. and Kellner, D. (2001) *The Postmodern Adventure: Science, Technology, and Cultural Studies at the Third Millennium*. London: Routledge.

Bijker, W. (1995) *Of Bicycles, Bakelite, and Bulbs*. Cambridge MA: MIT Press.

Bijker, W. and Law, J. (eds) (1992) *Shaping Technology/Building Society*. Cambridge MA: MIT Press.

Binkley, S. (2003) The seers of Menlo Park: the discourse of heroic consumption in the 'Whole Earth Catalog', *Journal of Consumer Culture*, 3(3): 283–313.

Bloor, D. (1991) *Knowledge and Social Imagery* 2nd edition. London: Routledge.

Bonner, F. (2003) *Ordinary Television*. London: Sage.

Bourdieu, P. (1984) *Distinction: A Social Critique of the Judgement of Taste*. London: Routledge.

Bozeman, T. (1997) Technological millenarianism in the United States, in T. Robbins and S. Palmer (eds) *Millennium, Messiahs, and Mayhem: Contemporary Apocalyptic Movements*. London: Routledge.

Brok, P. (2005) *Popular Science*. Buckingham: Open University Press.

Brosterman, N. (2000) *Out of Time: Designs for the Twentieth-century Future*. New York: Harry N. Abrams, Inc.

Brunsdon, C. (1997) *Screen Tastes: Soap Opera to Satellite*. London: Routledge.

Bryan, C. (1995) *Close Encounters of the Fourth Kind*. London: Orion.

Bryson, B. (2004) *A Short History of Nearly Everything*. London: Black Swan.

Bucchi, M. (2004) *Science in Society: An Introduction to Social Studies of Science.* London: Routledge.

Bukatman, S. (1995) The artificial infinite: on special effects and the sublime, in L. Cooke and P. Wollen (eds) *Visual Display: Culture Beyond Appearances.* Seattle WA: Bay Press.

Bukatman, S. (1997) *Blade Runner.* London: BFI.

Callinicos, A. (1989) *Against Postmodernism: A Marxist Critique.* Cambridge: Polity.

Callon, M. (1986) Some elements in a sociology of translation: domestication of the scallops and fishermen of St Brieuc Bay, in J. Law (ed.) *Power, Action and Belief.* London: Routledge & Kegan Paul.

Cavelos, J. (1998) *The Science of the X-Files.* New York: Berkeley.

Cavelos, J. (1999) *The Science of Star Wars.* New York: St Martin's Press.

Ceruzzi, P. (1998) *A History of Modern Computing.* Cambridge MA: MIT Press.

Clough, P. (2001) On the relationship of the criticism of ethnographic writing and the cultural studies of science, *Cultural Studies – Critical Methodologies*, 1(2): 240–70.

Cockburn, C. (1995) Black & Decker versus Moulinex, in S. Jackson and S. Moores (eds) *The Politics of Domestic Consumption: Critical Readings.* London: Prentice Hall.

Cocroft, W. and Thomas, R. (2003) *Cold War: Building for Nuclear Confrontation 1946–1989.* Swindon: English Heritage.

Coleman (pseudo.) (2003) I loathe Power Point, http: //www.eventweb.com

Collins, H. and Pinch, T. (1982) *Frames of Meaning: The Social Construction of Extraordinary Science.* London: Routledge and Kegan Paul.

Collins, J., Radner, H. and Preacher Collins, A. (eds) (1993) *Film Theory Goes to the Movies.* London: Routledge.

Collins, P. (2003) The future of lunar tourism, paper presented at the International Lunar Conference, Waikoloa, Hawaii, November; available from http:// www.spacefuture.com/pr/archive.shtml

Cook, J. (1999) Adapting telefantasy: the *Doctor Who and the Daleks* films, in I. Hunter (ed.) *British Science Fiction Cinema.* London: Routledge.

Cooper, R. and Parker, M. (1998) Cyborganization: cinema as nervous system, in J. Hassard and R. Holliday (eds) *Organization/Representation: Work and Organizations in Popular Culture.* London: Sage.

Cowan, R. (1985) How the refrigerator got its hum, in D. Mackenzie and J. Wajcman (eds) *The Social Shaping of Technology* 1st edition. Buckingham: Open University Press.

Creed, B. (1990) Alien and the monstrous-feminine, in A. Kuhn (ed.) *Alien Zone: Cultural Theory and Contemporary Science Fiction Cinema.* London: Verso.

Creed, B. (1993) *The Monstrous-Feminine: Film, Feminism, Psychoanalysis.* London: Routledge.

Crook, S., Pakulski, J. and Waters, M. (1992) *Postmodernization: Change in Advanced Society.* London: Sage.

Cubitt, S. (1999) Introduction. Le réel, c'est l'impossible: the sublime time of special effects, *Screen*, 40(2): 123–31.

Cull, N. (2001) 'Bigger on the inside . . .' *Doctor Who* as British cultural history, in G. Roberts and P. Taylor (eds) *The Historian, Television and Television History*. Luton: University of Luton Press.

Davis, E. (1998) *TechGnosis: Myth, Magic and Mysticism in the Age of Information*. London: Serpent's Tail.

Davis, K. (1995) *Reshaping the Female Body: The Dilemmas of Cosmetic Surgery*. London: Routledge.

Davis, K. (2003) *Dubious Equalities: Cultural Studies on Cosmetic Surgery*. Oxford: Rowman and Littlefield.

Davis, M. (1993) Dead West: ecocide in Marlboro country, *New Left Review*, 200: 49–73.

Dean, J. (1997) The truth is out there: aliens and the fugivity of postmodern truth, *Camera Obscura*, 40–41: 43–76.

Dean, J. (1998) *Aliens in America: Conspiracy Cultures from Outerspace to Cyberspace*. Ithaca NY: Cornell University Press.

Dery, M. (1996) *Escape Velocity: Cyberculture at the End of the Century*. London: Hodder & Stoughton.

Desmond, A. (1975) *The Hot-Blooded Dinosaur: A Revolution in Palaeontology*. London: Blond & Briggs.

Dixon, D. (2005) PSI wars: exploring 'Fortean Geographies' with the Mothman, paper presented at the Annual Meeting of the Association of American Geographers, Denver, March.

Du Gay, P., Hall, S., Janes, L, Mackay, H. and Negus, K. (1997) *Doing Cultural Studies: The Story of the Sony Walkman*. London: Sage.

Duffett, M. (2003) Imagined memories: webcasting as a 'live' technology and the case of Little Big Gig, *Information, Communication & Society*, 6(3): 307–25.

Durant, A. (1990) A new day for music?, in P. Hayward (ed.) *Culture, Technology and Creativity in the Late Twentieth Century*. London: John Libbey.

Edensor, T. (2002) *National Identity, Popular Culture and Everyday Life*. Oxford: Berg.

Edensor, T. (2005) *Industrial Ruins: Space, Aesthetics and Materiality*. Oxford: Berg.

Edgerton, G. (2004) The Moon landing, in G. Creeber (ed.) *Fifty Key Television Programmes*. London: Arnold.

Edwards, W. (1976) *The Early History of Palaeontology*. London: Natural History Museum.

Eisen, J. (1999) *Suppressed Inventions and Other Discoveries*. New York: Avery.

Elliott, D. (1982) The rock music industry, in Open University (eds) *Science, Technology and Popular Culture*. Milton Keynes: Open University Press.

Erikson, T. (2001) *Tyranny of the Moment: Fast and Slow in the Information Age*. London: Pluto.

Fahey, J., Bullen, E. and Kenway, J. (2005) A taste for science: inventing the young in the national interest, in D. Bell and J. Hollows (eds) *Ordinary Lifestyles: Popular Media, Consumption and Taste*. Buckingham: Open University Press.

Farish, M. (2003) Disaster and decentralization: American cities and the Cold War, *Cultural Geographies*, 10(2): 125–48.

Featherstone, M. (1991) *Postmodernism and Consumer Culture*. London: Sage.

Featherstone, M. (2002) *Knowledge and the Production of Nonknowledge. An Exploration of Alien Mythology in Post-war America*. Cresskill NJ: Hampton Press.

Felski, R. (2000) The invention of everyday life. *New Formations*, 39: 15–31.

Fenster, M. (1999) *Conspiracy Theories: Secrecy and Power in American Culture*. Minneapolis MN: University of Minnesota Press.

Fewer, G. (2002) Towards an LSMR and MSMR (Lunar and Martian Sites and Monuments Records): recording the planetary spacecraft landing sites as archaeological monuments for the future, in M. Russell (ed.) *Digging Holes in Popular Culture: Archaeology and Science Fiction*. Oxford: Oxbow Books.

Finn, C. (2001) *Artifacts: An Archaeologist's Year in Silicon Valley*. Cambridge MA: MIT Press.

Fisher, L. (2002) *How to Dunk a Doughnut: the Science of Everyday Life*. New York: Penguin.

Flicker, E. (2003) Between brains and breasts – women scientists in fiction film: on the marginalization and sexualization of scientific competence, *Public Understanding of Science*, 12(3): 307–18.

Foucault, M. (1975) *The Birth of the Clinic: An Archaeology of Medical Perception*. London: Routledge & Kegan Paul.

Foucault, M. (1979) *The History of Sexuality, Volume One: An Introduction*. Harmondsworth: Penguin.

Foucault, M. (2000) *Essential Works of Foucault 1954–1984: Ethics*, ed. P. Rabinow. London: Penguin.

Frith, S. (1986) Art versus technology: the strange case of popular music, *Media, Culture and Society*, 8(3): 263–80.

Frith, S. (1987) The industrialization of music, in J. Lull (ed.) *Popular Music and Communication*. London: Sage.

Fuller, M. (2003) *Behind the Blip: Essays on the Culture of Software*. New York: Autonomedia.

Fuller, S. (1997) *Science*. Buckingham: Open University Press.

Fuller, S. (2000) Science studies through the looking glass: an intellectual itinerary, in U. Segerstråle (ed.) *Beyond the Science Wars: The Missing Discourse about Science and Society*. New York: SUNY Press.

Gates, B. (1995) *The Road Ahead*. New York: Penguin.

Gauntlett, D. (2002) *Media, Gender and Identity: An Introduction*. London: Routledge.

Gauquelin, M. (1983) *The Truth about Astrology*. Oxford: Blackwell.

Giddens, A. (1991) *Modernity and Self-identity*. Cambridge: Polity.

Gieryn, T. (1995) The boundaries of science, in S. Jasanoff, G. Markle, J. Peterson and T. Pinch (eds) *Handbook of Science and Technology Studies*. London: Sage.

Gieryn, T. (1998) Balancing acts: science, *Enola Gay* and History Wars at the Smithsonian, in S. Macdonald (ed.) *The Politics of Display: Museums, Science, Culture*. London: Routledge.

Gieryn, T. (1999) *Cultural Boundaries of Science: Credibility on the Line*. Chicago: University of Chicago Press.

Goldman, R. and Papson, S. (1996) *Sign Wars: The Cluttered Landscape of Advertising*. London: Guilford.

Goodwin, C. (1995) Seeing in depth, *Social Studies of Science*, 25: 237–74.

Gottdeiner, M., Collins, C. and Dickens, D. (1999) *Las Vegas: The Social Production of an All-American City*. Oxford: Blackwell.

Gray, A. (1995) Technology in the domestic environment, in S. Jackson and S. Moores (eds) *The Politics of Domestic Consumption: Critical Readings*. London: Prentice Hall.

Gregory, J. (2003) The popularization and excommunication of Fred Hoyle's 'life-from-space' theory, *Public Understanding of Science*, 12(1): 25–46.

Gregory, J. and Miller, S. (1998) *Science in Public: Communication, Culture, and Credibility*. Cambridge MA: Basic Books.

Gregory, J. and Miller, S. (2000) *Science in Public: Communication, Culture, and Credibility*. Cambridge MA: Basic Books.

Gresh, L. and Weinberg, R. (2002) *The Science of Superheroes*. London: Wiley.

Gross, P. and Levitt, N. (1994) *Higher Superstition: The Academic Left and its Quarrels with Science*. Baltimore MD: Johns Hopkins University Press.

Gusterson, H. (2004a) Nuclear tourism, *Journal for Cultural Research*, 8(1): 23–30.

Gusterson, H. (2004b) *People of the Bomb: Portraits of America's Nuclear Complex*. Minneapolis MN: University of Minnesota Press.

Hankins, J. (2004) Lost in space, *Observer Magazine*, 20 March: 59.

Haraway, D. (1991) *Simians, Cyborgs, and Women: The Reinvention of Nature*. London: Free Association Books

Haraway, D. (1997) *Modest_Witness@Second_Millennium.FemaleMan©_Meets_-OncoMouse™*. London: Routledge.

Haraway, D. (2003) *The Companion Species Manifesto: Dogs, People, and Significant Otherness*. Chicago: Prickly Paradigm Press.

Harrison, D., Pile, S. and Thrift, N. (2004) *Patterned Ground: Entanglements of Nature and Culture*. London: Reaktion.

Hawking, S. (1988) *A Brief History of Time: From the Big Bang to Black Holes*. London: Bantam.

Hayles, N.K. (1996) Consolidating the canon, in A. Ross (ed.) *Science Wars*. Durham NC: Duke University Press.

Haynes, R. (1994) *From Faust to Strangelove: Representations of the Scientists in Western Literature*. Baltimore MD: Johns Hopkins University Press.

Haynes, R. (2003) From alchemy to artificial intelligence: stereotypes of the scientist in Western literature, *Public Understanding of Science*, 12(3): 243–53.

Henrikson, M. (1997) *Dr Strangelove's America: Society and Culture in the Atomic Age*. Berkeley CA: University of California Press.

Hess, D. (1993) *Science in the New Age: The Paranormal, its Defenders and Debunkers, and American Culture*. Madison WI: University of Wisconsin Press.

Hetherington, K. (2004) Secondhandedness: consumption, disposal and absent presence, *Environment & Planning D: Society & Space*, 22(1): 157–73.

Higgins, K. (2001) *Collecting the 1970s*. London: Miller's.

Highfield, R. (2002) *The Science of Harry Potter*. New York: Viking.

Highmore, B. (2002a) *Everyday Life and Cultural Theory*. London: Routledge.

Highmore, B. (2002b) *The Everyday Life Reader*. London: Routledge.

Higley, S. (1997) Alien intellect and the roboticization of the scientist, *Camera Obscura*, 40–41: 131–62.

Hilgartner, S. (2000) *Science on Stage: Expert Advice as Public Drama*. Stanford CA: Stanford University Press.

Holliday, R. (2005a) Home truths? In D. Bell and J. Hollows (eds) *Ordinary Lifestyles: Popular Media, Consumption and Taste*. Buckingham: Open University Press.

Holliday, R. (2005b) Review essay, *Dubious Equalities* and *In The Flesh*, *Sociology of Health and Illness*, 27(1): 157–60.

Holliday R. and Sanchez Taylor, J. (2006) Aesthetic surgery as false beauty, *Feminist Theory*, forthcoming.

Holliday, R. and Potts, T. (2006) *Kitsch: A Cultural Politics of Taste*. Manchester: Manchester University Press.

Holt-Jensen, A. (1980) *Geography: Its History and Concepts*. London: Sage.

Holton, G. (1993) *Science and Anti-science*. Cambridge MA: Harvard University Press.

Hoyle, F. and Wickramasinghe, C. (1983) *Evolution from Space*. London: Paladin.

Hoyle, F. and Wickramasinghe, C. (1986) *Archaeopteryx, the Primordial Bird: A Case of Fossil Forgery?* London: Christopher Davies.

Humphery, K. (1998) *Shelf Life: Supermarkets and the Changing Cultures of Consumption*. Cambridge: Cambridge University Press.

Hunter, I. (1999) Introduction: the strange world of the British science fiction film, in I. Hunter (ed.) *British Science Fiction Cinema*. London: Routledge.

Huyghe, P. (1996) *The Field Guide to Extraterrestrials*. London: Hodder & Stoughton.

Irwin, A. and Michael, M. (2003) *Science, Social Theory and Public Knowledge*. Maidenhead: Open University Press.

Irwin, A., Allan, S. and Welsh, I. (2000) Nuclear risks: three problematics, in B. Adam, U. Bech and J. van Loon (eds) *The Risk Society and Beyond: Critical Issues for Social Theory*. London: Sage.

Jackson, S. and Moores, S. (eds) (1995) *The Politics of Domestic Consumption: Critical Readings*. London: Prentice Hall.

Johnston, R. *et al.* (eds) (2000) *The Dictionary of Human Geography* 4th edition. Oxford: Blackwell.

Jones, R. (1997) The Boffin: a stereotype of scientists in post-war British films (1945–1970), *Public Understanding of Science*, 6(1): 31–48.

Jones, R. (1998) The scientists as artist: a study of *The Man in the White Suit* and some related British film comedies of the postwar period (1945–1970), *Public Understanding of Science*, 7(3): 135–47.

Jones, R. (2001) 'Why can't you scientists leave things alone?' Science questioned in

British films of the post-war period (1945–1970), *Public Understanding of Science*, 10(4): 365–82.

Jorg, D. (2003) The good, the bad and the ugly – Dr Moreau goes to Hollywood, *Public Understanding of Science*, 12(3): 297–305.

Kaplan, E. (1987) *Rocking Around the Clock: Music Television, Postmodernism, and Consumer Culture*. London: Methuen.

Keller, J. (2003) PowerPoint populace, *TwinCities.com* http://www.twincities.com (accessed 22 June 2005).

Kennedy, B. (2000) *Deleuze and Cinema: The Aesthetics of Sensation*. Edinburgh: Edinburgh University Press.

Kiernan, V. (2000) The Mars meteorite: a case study in controls on dissemination of news, *Public Understanding of Science*, 9(1): 15–41.

King, L. (ed.) (2002) *Game On: The History and Culture of Videogames*. London: Laurence King.

Kirsch, S. (1997) Watching the bombs go off: photography, nuclear landscapes, and spectacular democracy, *Antipode*, 29(3): 227–55.

Kirsch, S. (2000) Peaceful nuclear explosions and the geography of scientific authority, *Professional Geographer*, 52(2): 179–92.

Kline, S. (2003) What is technology? In R. Scharff and V. Dusek (eds) *Philosophy of Technology: The Technological Condition*. Oxford: Blackwell.

Knight, P. (2000) *Conspiracy Culture: From Kennedy to The X-Files*. London: Routledge.

Knorr-Cetina, K. (1981) *The Manufacture of Knowledge: An Essay on the Constructivist and Contextual Nature of Science*. Oxford: Pergamon.

Kozlovic, A. (2003) Technophobic themes in pre-1990 computer films, *Science as Culture*, 12(3): 341–73.

Krauss, L. (1995) *The Physics of Star Trek*. London: Flamingo.

Kuhn, A. (1990) Introduction: cultural theory and science fiction cinema, in A. Kuhn (ed.) *Alien Zone: Cultural Theory and Science Fiction Cinema*. London: Verso.

Kuhn, A. (ed) (1999) *Alien Zone II: The Spaces of Science Fiction Cinema*. London: Verso.

Kuhn, T. (1962) *The Structure of Scientific Revolutions*. Chicago: Chicago University Press.

Lacquer, T. (1990) *Making Sex: Body and Gender from the Greeks to Freud*. Cambridge MA: Harvard University Press.

Laing, G. (2004) *Digital Retro: The Evolution and Design of the Personal Computer*. Lewes: Ilex.

Lally, E. (2002) *At Home with Computers*. Oxford: Berg.

Lambourne, R. (1999) Science fiction and the communication of science, in E. Scanlon, E. Whitelegg and S. Yates (eds) *Communicating Science: Contexts and Channels*. London: Routledge.

Lang, D. (1952/1995) Blackjack and flashes, in M. Tronnes (ed.) *Literary Las Vegas: The Best Writing about America's Most Fabulous City*. New York: Henry Holt.

Lantry, D. (2001) Dress for egress: the Smithsonian National Air and Space Museum's Apollo spacesuit collection, *Journal of Design History*, 14(4): 343–59.

Latour, B. (1987) *Science in Action*. Milton Keynes: Open University Press.

Latour, B. (1992) Where are the missing masses? A sociology of a few mundane artifacts, in W. Bijker and J. Law (eds) *Shaping Technology/Building Society*. Cambridge MA: MIT Press.

Latour, B. (1993) *We Have Never Been Modern*. Hemel Hempstead: Harvester Wheatsheaf.

Latour, B. and Woolgar, S. (1979) *Laboratory Life: The Social Construction of Scientific Facts*. London: Sage.

Law, J. (1987) Technology and heterogeneous engineering: the case of Portuguese expansion, in W. Bijker, T. Hughes and T. Pinch (eds) *Social Construction of Technological Systems*. Cambridge MA: MIT Press.

Law, J. (1999) After ANT: complexity, naming and topology, in J. Law and J. Hassard (eds) *Actor Network Theory and After*. Oxford: Blackwell.

Law, J. and Hassard, J. (eds) (1999) *Actor Network Theory and After*. Oxford: Blackwell.

Leary, T. (1994) *Chaos and Cyberculture*. Berkeley CA: Ronin.

Lehtonen, T-K. (2003) The domestication of new technologies as a set of trials, *Journal of Consumer Culture*, 3(3): 363–85.

Leib, M. (1998) *Children of Ezekiel: Aliens, UFOs, the Crisis of Race, and the Advent of End Time*. Durham NC: Duke University Press.

LeVay, S. (1993) *The Sexual Brain*. Cambridge MA: MIT Press.

Levine, G. (1996) What is science studies for and who cares? in A. Ross (ed.) *Science Wars*. Durham NC: Duke University Press.

Leyshon, A. (2001) Time-space (and digital) compression: software formats, musical networks, and the reorganization of the music industry, *Environment & Planning A:* 33(1): 49–77.

Leyshon, A. (2003) Scary monsters? Software formats, peer-to-peer networks, and the spectre of the gift, *Environment & Planning D: Society & Space*, 21(6): 533–58.

Lifton, R. and Markusen, E. (1990) *The Genocidal Mentality*. New York: Basic Books.

Light, M. (1999) *Full Moon*. London: Jonathan Cape.

Light, M. (2003) *100 Suns*. London: Jonathan Cape.

Loader, B. (ed.) (1998) *Cyberspace Divide: Equality, Agency and Policy in the Information Society*. London: Routledge.

Lupton, D. and Noble, G. (1997) Just a machine? Dehumanizing strategies in personal computer use, *Body & Society*, 3(1): 83–101.

Lupton, D. and Noble, G. (2002) Mine/not mine: appropriating personal computers in the academic workplace, *Journal of Sociology*, 38(1): 5–23.

Lyotard, J-F. (1984) *The Postmodern Condition: A Report on Knowledge*. Manchester: Manchester University Press.

Mackay, H. (1997) Consuming communication technologies at home, in H. Mackay (ed.) *Consumption and Everyday Life*. London: Sage.

Mackenzie, D. and Wajcman, J. (1999) Introductory essay and general issues, in D. Mackenzie and J. Wajcman (eds) *The Social Shaping of Technology* 2nd edition. Buckingham: Open University Press.

McCurdy, H. (1997) *Space and the American Imagination*. Washington: Smithsonian Institution Press.

McGuigan, J. (1992) *Cultural Populism*. London: Routledge.

McLeod, J. (1997) *Narrative and Psychotherapy*. London: Sage.

Michael, M. (1996) *Constructing Identities*. London: Sage.

Michael, M. (2000) *Reconnecting Culture, Technology and Nature: From Society to Heterogeneity*. London: Routledge.

Michael, M. (2003) Between the mundane and the exotic: time for a different socio-technical stuff, *Time & Society*, 12(1): 127–43.

Miller, D. and Slater, D. (2000) *The Internet: An Ethnographic Approach*. Oxford: Berg.

Milton, K. (ed.) (1993) *Environmentalism: The View from Anthropology*. London: Routledge.

Mitchell, W. (1998) *The Last Dinosaur Book: The Life and Times of a Cultural Icon*. Chicago: University of Chicago Press.

Molella, A. (2003) Exhibiting atomic culture: the view from Oak Ridge, *History and Technology*, 19(3): 211–26.

Molotch, H. (2003) *Where Stuff Comes From: How Toasters, Toilets, Cars, Computers, and Many Other Things Come to Be as They Are*. London: Routledge.

Morgan, K. (2003) Women and the knife: cosmetic surgery and the colonization of women's bodies, in R. Weitz (ed.) *The Politics of Women's Bodies: Sexuality, Appearance, and Behavior* 2nd edition. Oxford: Oxford University Press.

Mulholland, G. (2002) *This is Uncool*. London: Cassell.

Murphie, A. and Potts, J. (2003) *Culture and Technology*. Basingstoke: Palgrave Macmillan.

Murphy, T. (1997) *Gay Science: The Ethics of Sexual Orientation Research*. New York: Columbia University Press.

Nader, L. (1996) Introduction: anthropological inquiry into boundaries, power, and knowledge, in L. Nader (ed.) *Naked Science: Anthropological Inquiry into Boundaries, Power, and Knowledge*. London: Routledge.

Nader, L. (ed.) (1996) *Naked Science: Anthropological Inquiry into Boundaries, Power, and Knowledge*. London: Routledge.

Negus, K. (1996) *Popular Music in Theory*. Cambridge: Polity.

Newkey-Burden, C. (2003) *Nuclear Paranoia*. Harpenden: Pocket Essentials.

Noble, D. (1984) *Forces of Production*. New York: Alfred Knopf.

Nye, D. (1994) *American Technological Sublime*. Cambridge MA: MIT Press.

O'Neill, J. (1996) Dinosaurs-R-us: the (un)natural history of *Jurassic Park*, in J. Cohen (ed.) *Monster Theory: Reading Culture*. Minneapolis MN: University of Minnesota Press.

O'Sullivan, T. (2005) Television and the framing and classification of lifestyles, in D. Bell and J. Hollows (eds) *Ordinary Lifestyles: Popular Media, Consumption and Taste*. Buckingham: Open University Press.

Oksanen-Sarela, K. and Pantzar, M. (2001) Smart life, version 3.0: representations of everyday life in future studies, in J. Gronow and A. Warde (eds) *Ordinary Consumption*. London: Routledge.

Ormrod, S. (1994) 'Let's nuke the dinner': discursive practices of gender in the creation of a new cooking process, in C. Cockburn and R. Fürst-Dilić (eds) *Bringing Technology Home: Gender and Technology in a Changing Europe*. Buckingham: Open University Press.

Palen, L., Salzman, M. and Youngs, E. (2001) Discovery and integration of mobile communications in everyday life, *Personal and Ubiquitous Computing*, 5(2): 109–22.

Paradis, T. (2002) The political economy of theme development in small urban places: the case of Roswell, New Mexico, *Tourism Geographies*, 4(1): 22–43.

Parkins, W. and Craig, G. (2005) *Slow Living*. Oxford: Berg.

Penley, C. (1997) *NASA/TREK: Popular Science and Sex in America*. London: Verso.

Penley, C. and Ross, A. (1991) Introduction, in C. Penley and A. Ross (eds) *Technoculture*. Minneapolis MN: University of Minnesota Press.

Pickover, C. (1999) *The Science of Aliens*. New York: Basic Books.

Pyle, F. (2000) Making cyborgs, making humans: of Terminators and Blade Runners, in D. Bell and B. Kennedy (eds) *The Cybercultures Reader*. London: Routledge.

Raab, T. and Frodeman, R. (2002) What is it like to be a geologist? A phenomenology of geology and its epistemological implications, *Philosophy and Geography*, 5(1): 69–81.

Ramsland, K. (2001) *The Forensic Science of CSI*. New York: Berkeley.

Rathje, W. (1999) Archaeology of space garbage, *Discovering Archaeology*, 1(3): 108–11.

Redclift, M. and Benton, T. (eds) (1994) *Social Theory and the Global Environment*. London: Routledge.

Reid, R. and Traweek, S. (eds) (2000) *Doing Science + Culture: How Cultural and Interdisciplinary Studies are Changing the Way We Look at Science and Medicine*. London: Routledge.

Reinel, B. (1999) Reflections on cultural studies of technoscience, *European Journal of Cultural Studies*, 2(2): 163–89.

Restivo, S. (2001) 4S, the FBI, and anarchy, *Science, Technology & Human Values*, 26(1): 87–90.

Rheingold, H. (1993) *Virtual Community: Homesteading on the Electronic Frontier*. Reading MA: Addison Wesley.

Robins, K. and Webster, F. (1999) *Times of the Technoculture: From the Information Society to the Virtual Life*. London: Routledge.

Rojek, C. (2001) *Leisure and Culture*. Basingstoke: Macmillan.

Rosen, P. (1997) 'It was easy, it was cheap, go and do it!' Technology and anarchy in the UK music industry, in J. Purkis and J. Bowen (eds) *Twenty-First Century Anarchism*. London: Cassell.

Rosen, P. (2002) *Framing Production: Technology, Culture, and Change in the British Bicycle Industry*. Cambridge MA: MIT Press.

Ross, A. (1991) *Strange Weather: Culture, Science and Technology in the Age of Limits*. London: Verso.

Ross, A. (1994) The new smartness, in G. Bender and T. Druckrey (eds) *Culture on the Brink: Ideologies of Technology*. Seattle WA: Bay Press.

Ross, A. (ed.) (1996) *Science Wars*. Durham: Duke University Press.

Ryan, M. and Kellner, D. (1990) Technophobia, in A. Kuhn (ed.) *Alien Zone: Cultural Theory and Science Fiction Cinema*. London: Verso.

Sagan, C. (1997) *The Demon-haunted World: Science as a Candle in the Dark*. London: Headline.

Saks, M. (2003) *Orthodox and Alternative Medicine*. London: Continuum.

Sardar, Z. (2000) *Thomas Kuhn and the Science Wars*. Cambridge: Icon.

Sassower, R. (1995) *Cultural Collisions: Postmodern Technoscience*. London: Routledge.

Segerstråle, U. (2000) Anti-antiscience: a phenomenon in search of an explanation, in U. Segerstråle (ed.) *Beyond the Science Wars: The Missing Discourse about Science and Society*. New York: SUNY Press.

Segerstråle, U. (ed.) (2000) *Beyond the Science Wars: The Missing Discourse about Science and Society*. New York: SUNY Press.

Self, W. (1995) *Scale*. Harmondsworth: Penguin.

Shawyer, L. (n.d.) Postmodern Therapies News, http://www.california.com/~rathbone/pmth.htm

Shove, E. (2003) *Comfort, Cleanliness and Convenience: The Social Organization of Normality*. Oxford: Berg.

Shove, E. and Southerton, D. (2000) Defrosting the freezer: from novelty to convenience, *Journal of Material Culture*, 5(3): 301–19.

Sim, S. (2002) *Irony and Crisis: A Critical History of Postmodern Culture*. Cambridge: Icon.

Simon, A. (2001) *The Real Science Behind The X-Files*. New York: Simon & Schuster.

Sismondo, S. (2004) *An Introduction to Science and Technology Studies*. Oxford: Blackwell.

Snow, C. (1963) *The Two Cultures*. Cambridge: Cambridge University Press.

Sokal, A. (1996a) Transgressing the boundaries: toward a transformative hermeneutics of quantum gravity, *Social Text*, 46–7: 217–52.

Sokal, A. (1996b) A physicist experiments with cultural studies, *Lingua Franca*, 6(4): 62–4.

Sokal, A. and Bricmont, J. (1997) *Intellectual Impostures*. London: Profile Books.

Spigel, L. (2001) Media homes, then and now, *International Journal of Cultural Studies*, 4(4): 385–411.

Spufford, F. (2003) *Backroom Boys: The Secret Return of the British Boffin*. London: Faber & Faber.

Squeri, L. (2004) When ET calls: SETI is ready, *Journal of Popular Culture*, 37(4): 478–96.

Star, S. (1999) The ethnography of infrastructure, *American Behavioral Scientist*, 43(3): 377–91.

Stern, M. (2003) A brief history of Stephen Hawking: making scientific meaning in contemporary Anglo-American culture, *New Formations*, 50: 150–64.

Sterne, J. (1999) Thinking the Internet: cultural studies versus the millennium, in S. Jones (ed.) *Doing Internet Research: Critical Issues and Methods for Examining the Net*. London: Sage.

Stivers, R. (2001) *Technology as Magic: The Triumph of the Irrational*. London: Continuum.

Straw, W. (1997) Sizing up record collections: gender and connoisseurship in rock music culture, in S. Whiteley (ed.) *Sexing the Groove: Popular Music and Gender*. London: Routledge.

Strum, W. (2001) We were the Unabomber, *Science, Technology & Human Values*, 26(1): 90–101.

Tapia, A. (2003) Technomillennialism: a subcultural response to the technological threat of Y2K, *Science, Technology & Human Values*, 28(4): 483–512.

Taylor, P. (1999) *Hackers: Crime in the Digital Sublime*. London: Routledge.

Taylor, T. (2001) *Strange Sounds: Music, Technology and Culture*. London: Routledge.

Tenner, E. (1996) *When Things Bite Back: Predicting the Problems of Progress*. London: Fourth Estate.

Terranova, T. (2000) Post-human unbounded: artificial evolution and high-tech sub-cultures, in D. Bell and B. Kennedy (eds) *The Cybercultures Reader*. London: Routledge.

Thieme, R. (2000) Stalking the UFO meme, in D. Bell and B. Kennedy (eds) *The Cybercultures Reader*. London: Routledge.

Thompson, D. (2004) Exposure, *Eye*, 51: 18–27.

Thompson, G. (2004) *The Business of America: The Cultural Production of a Post-war Nation*. London: Pluto.

Thorburn, G. (2002) *Men and Sheds*. London: New Holland.

Thwaites, T., Davis, L. and Mules, W. (1994) *Tools for Cultural Studies*. Melbourne: Macmillan.

Tiefer, E. (1995) *Sex is Not a Natural Act, and Other Essays*. Boulder CA: Westview.

Toop, D. (1995) *Oceans of Sound: Aether Talk, Ambient Sound and Imaginary Worlds*. London: Serpent's Tail.

Traweek, S. (1988) *Beamtimes and Lifetimes: The World of High Energy Physicists*. Cambridge MA: Harvard University Press.

Trocco, F. (1998) How to believe in weird things, *Public Understanding of Science*, 7(1): 187–93.

Tudor, A. (1989) *Monsters and Mad Scientists: A Cultural History of the Horror Movie*. Oxford: Blackwell.

Tulloch, J. and Jenkins, H. (1995) *Science Fiction Audiences: Watching Doctor Who and Star Trek*. London: Routledge.

Tully, C. (2003) Growing up in technological worlds: how modern technologies shape the everyday lives of young people, *Bulletin of Science, Technology and Society*, 23(6): 444–56.

Turkle, S. (2003) From powerful ideas to PowerPoint, *Convergence*, 9(2): 19–25.

Turney, J. (1998) *Frankenstein's Footsteps: Science, Genetics and Popular Culture*. New Haven CT: Yale University Press.

Ullman, E. (1997) *Close to the Machine*. San Francisco: City Lights.

Vanderbilt, T. (2002) *Survival City: Adventures among the Ruins of Atomic America*. Princeton NJ: Princeton Architectural Press.

Verren, H. (2001) *Science and an African Logic*. Chicago: University of Chicago Press.

Virilio, P. (1989) *The Museum of Accidents*. Toronto: Public Access Collective.

Virilio, P. (1994) *Bunker Archaeology*. Princeton NJ: Princeton Architectural Press.

Wagar, W. (1995) The mad bad scientist, *Science Fiction Studies*, 22(1): 1–5.

Wajcman, J. (1995) Domestic technology: labour-saving or enslaving?, in S. Jackson and S. Moores (eds) *The Politics of Domestic Consumption: Critical Readings*. London: Prentice Hall.

Waldby, C. (2000) *The Visible Human Project: Informatic Bodies and Posthuman Medicine*. London: Routledge.

Warner, M. (2002) *Publics and Counterpublics*. New York: Zone Books.

Webster, D. (1988) *Looka Yonder! The Imaginary America of Populist Culture*. London: Routledge.

Weingart, P. (2003) Of power maniacs and unethical geniuses: science and scientists in fiction film, *Public Understanding of Science*, 12(3): 279–87.

Weintraub, I. (2000) The impact of alternative presses on scientific communication, *International Journal on Grey Literature*, 1(2): 54–9.

Wilcox, R. and Williams, J. (1996) 'What do you think?' *The X-Files*, liminality, and gender pleasure, in D. Lavery, A. Hague and M. Cartwright (eds) *Deny All Knowledge: Reading the X-Files*. London: Faber.

Williams, M. (2000) *Science and Social Science: An Introduction*. London: Routledge.

Willis, R. and Curry, P. (2004) *Astrology, Science and Culture: Pulling Down the Moon*. Oxford: Berg.

Wills, J. (2003) Abalone, rattlesnakes and kilowatt monsters: nature and the atom at Diablo Canyon, California, *Cultural Geographies*, 10(2): 149–175.

Wood, A. (2002) *Technoscience in Contemporary American Film: Beyond Science Fiction*. Manchester: Manchester University Press.

Woodward, R. (2004) *Military Geographies*. Oxford: Blackwell.

Woolgar, S. (1988) *Science: The Very Idea*. London: Routledge.

Yaco, L. and Haber, K. (2004) *The Science of the X-Men*. New York: Marvel.

Yearley, S. (2005) *Making Sense of Science: Understanding the Social Study of Science*. London: Sage.

Ziman, J. (1992) Not knowing, needing to know, and wanting to know, in B. Lewenstein (ed.) *When Science Meets the Public*. Washington: AAAS.

Žižek, S. (2001) Welcome to the desert of the real, *Re: Constructions*, http://web.mit-.edu/cms/reconstructions/interpretations/desertreal.html

INDEX

Related books from Open University Press
Purchase from www.openup.co.uk or order through your local bookseller

MEDIA, RISK AND SCIENCE
Stuart Allan

- How is science represented by the media?
- Who defines what counts as a risk, threat or hazard, and why?
- In what ways do media images of science shape public perceptions?
- What can cultural and media studies tell us about current scientific controversies?

Media, Risk and Science is an exciting exploration into an array of important issues, providing a much needed framework for understanding key debates on how the media represent science and risk. In a highly effective way, Stuart Allan weaves together insights from multiple strands of research across diverse disciplines. Among the themes he examines are: the role of science in science fiction, such as *Star Trek*; the problem of 'pseudo-science' in *The X-Files*; and how science is displayed in science museums. Science journalism receives particular attention, with the processes by which science is made 'newsworthy' unravelled for careful scrutiny. The book also includes individual chapters devoted to how the media portray environmental risks, HIV-AIDS, food scares (such as BSE or 'mad cow disease' and GM foods) and human cloning. The result is a highly topical text that will be invaluable for students and scholars in cultural and media studies, science studies, journalism, sociology and politics.

Contents
Series editor's foreword – Introduction: media, risk and science – Science fictions – Science in popular culture – Science journalism – Media, risk and the environment – Bodies at risk: news coverage of AIDS – Food scares: mad cows and GM foods – Figures of the human: robots, androids, cyborgs and clones – Glossary – References – Index.

256pp   0 335 20662 X (Paperback)   0 335 20663 8 (Hardback)

ORDINARY LIFESTYLES
POPULAR MEDIA, CONSUMPTION AND TASTE

David Bell and Joanne Hollows (eds)

Lifestyle media – books, magazines, websites, radio and television shows that focus on topics such as cookery, gardening, travel and home improvement – have witnessed an explosion in recent years.

Ordinary Lifestyles explores how popular media texts bring ideas about taste and fashion to consumers, helping audiences to fashion their lifestyles as well as defining what constitutes an appropriate lifestyle for particular social groups. Contemporary examples are used throughout, including Martha Stewart, *House Doctor, What Not to Wear, You Are What You Eat, Country Living* and brochures for gay and lesbian holiday promotions.

The contributors show that watching make-over television or cooking from a celebrity chef's book are significant cultural practices, through which we work on our ideas about taste, status and identity. In opening up the complex processes which shape our taste and forge individual and collective identities, lifestyle media demand our serious attention, as well as our viewing, reading and listening pleasure.

Ordinary Lifestyles is essential reading for students on media and cultural studies courses, and for anyone intrigued by the influence of the media on our day-to-day lives.

Contributors
David Bell, Frances Bonner, Steven Brown, Fan Carter, Stephen Duncombe, David Dunn, Johannah Fahey, Elizabeth Bullen, Jane Kenway, Robert Fish, Danielle Gallegos, Mark Gibson, David B. Goldstein, Ruth Holliday, Joanne Hollows, Felicity Newman, Tim O'Sullivan, Elspeth Probyn, Rachel Russell, Lisa Taylor, Melissa Tyler, Gregory Woods.

Contents
Ordinary Lifestyles – SECTION I: MEDIA FORM AND INDUSTRY – From Television Lifestyle to Lifestyle Television – Whose Lifestyle is it Anyway? – Recipes for Living: Martha Stewart and the New American Subject – SECTION II: HOME FRONT – Home Truths? – Monoculture versus Multiculinarism: Trouble in the Aussie Kitchen – Cookbooks as Manuals of Taste – SECTION III: THE GREAT OUTDOORS – It was Beautiful Before You Changed it All: Class, Taste and the Transformative Aesthetics of the Garden Lifestyle Media – Entertaining Tourists: Television Holiday Programmes, Performance, and the Tourist Destination – Holidays of a Lifestyle: Representations of Pleasure in Gay and Lesbian Holiday Promotions – Countryside Formats and Ordinary Lifestyles – SECTION IV: LEARNING LIFESTYLES – It's a Girl Thing: Teenage Magazines, Lifestyle and Consumer Culture – Gender, Childhood and Consumer Culture – A Taste for Science: Inventing the Young in the National Interest – SECTION V: WORK/LIFE BALANCING – Sabotage, Slack and the Zinester Search for Non-Alienated Labour – The Worst Things in the World: Life Events Checklists in Popular Stress – Management Texts – Thinking Habits and the Ordering of Life – Bibliography

224pp   0 335 21550 5 (Paperback)   0 335 21551 3 (Hardback)

WHAT IS THIS THING CALLED SCIENCE?
AN ASSESSMENT OF THE NATURE AND STATUS OF SCIENCE AND ITS METHODS

Alan Chalmers

Reviews of the previous edition:

> In this academic bestseller – indeed, one of the most widely read books ever written in the history and philosophy of science – Alan Chalmers provides a refreshingly lucid introduction ... Drawing on illuminating historical examples, he asks and answers some of the most fundamental questions about the nature of science and its methods.
>
> Ronald L. Numbers, William Coleman Professor of the History of Science
and Medicine, University of Wisconsin at Madison

> Crisp, lucid and studded with telling examples ... As a handy guide to recent alarums and excursions (in the philosophy of science) I find this book vigorous, gallant and useful.
>
> *New Scientist*

- What is the characteristic that serves to distinguish scientific knowledge from other kinds of knowledge?
- What is the role of experiment in science?
- What is the role of theory in science?

In clear, jargon-free language, the third edition of this highly successful introduction to the philosophy of science surveys the answers of the past hundred years to these central questions. This edition has been enriched by many new historical examples and the early chapters have been reorganised, re-ordered and amplified to facilitate the introduction of beginners to the field. The third edition includes new chapters on the new experimentalism; the Bayesian approach to science; the nature of scientific laws; and recent developments in the realism/anti-realism debate.

Contents

Preface – Introduction – Science as knowledge derived from the facts of experience – Observation as practical intervention – Experiment – Deriving theories from the facts: induction – Introducing falsificationism – Sophisticated falsificationism: novel predictions and the growth of science – The limitations of falsificationism – Theories as structures I: Kuhn's paradigms – Theories as structures II: Research programmes – Feyerabend's anarchistic theory of science – Methodical changes in method – The Bayesian approach – The new experimentalism – Why should the world obey laws? – Realism and anti-realism – Epilogue – Bibliography – Index.

288pp   0 335 20109 1 (Paperback)
Not available in Australia, Asia, South America and North America.

UNDERSTANDING POPULAR SCIENCE
Peter Broks

This is a lively book which provides a framework to help understand the development of popular science and current debates about it. In an easily accessible style, Peter Broks shows how popular science has been invented, redefined and fought over. From early-nineteenth century radical science to twenty-first century government initiatives, he examines popular science as an arena where the authority of science and the authority of the state are legitimated and challenged.

The book includes clear accounts of the public perception of scientists, visions of the future, the 'Two Cultures' debate and concerns about scientific literacy. The final provocative chapter proposes a new model for understanding the interaction between lay and expert knowledge.

Essential reading in cultural studies, science studies, history of science and science communication.

Contents
Introduction – Uncertain Times – The Rise of the Expert – Science and Modernity – Consuming Doubts – The Public Understanding of Science – From PUS to CUSP – Conceptual Space – Glossary – Bibliography – Index

192pp   0 335 21548 3 (Paperback)   0 335 21549 1 (Hardback)

DOMESTICATION OF MEDIA AND TECHNOLOGY

Maren Hartmann, Thomas Berker, Yves Punie and Katie Ward (eds)

This book provides an overview of a key concept in media and technology studies: domestication. Theories around domestication shed light upon the process in which a technology changes its status from outrageous novelty to an aspect of everyday life which is taken for granted. The contributors collect past, current and future applications of the concept of domestication, critically reflect on its theoretical legacy, and offer comments about further development.

The first part of *Domestication of Media and Technology* provides an overview of the conceptual development and theory of domestication. In the second part of the book, contributors look at a diverse range of empirical studies that use the domestication approach to examine the dynamics between users and technologies. These studies include:

- Mobile information and communications techologies (ICTs) and the transformation of the relationship between private and the public spheres
- Home-based internet use: the two-way dynamic between the household and its social environment
- Disadvantaged women in Europe undertaking introductory internet courses
- Urban middle-class families in China who embrace ICTs and view them as instruments of upward mobility and symbols of success

The book offers valuable insights for both experienced researchers and students looking for an introduction to the concept of domestication.

Contributors
Maria Bakardjieva, Thomas Berker, Leslie Haddon, Maren Hartmann, Deirdre Hynes, Sun Sun Lim, Anna Maria Russo Lemor, David Morley, Jo Pierson, Yves Punie, Els Rommes, Roger Silverstone, Knut H. Sørensen, Katie J. Ward.

Contents
Introduction – I. Theory and History – Exploring Domestication Today – Domestication: The social performance of technology – Empirical Studies Using the Domestication Framework – Domestication Run Wild: From the Moral Economy of the Household to the Mores of a Culture – And Where is the Content? Media as technological objects, symbolic environments and individual texts – II. Theory & Application – Fitting the internet into our lives: what IT courses have to do with it – Domestication, Home and Work – Making a 'Home': the domestication of information and communication technologies in Single Parents' households – From cultural to information revolution: ICT domestication by middle-class Chinese families – Domestication at work in small businesses – III. Summary – Domesticating Domestication?

2005   240pp   0 335 21768 0 (Paperback)   0 335 21769 9 (Hardback)